Dynamic Structures in an Open Cosmos

Stigmergy and Vortices

Mathias Hüfner

Dynamic Structures in an Open Cosmos

Stigmergy and Vortices

by

Mathias Hüfner

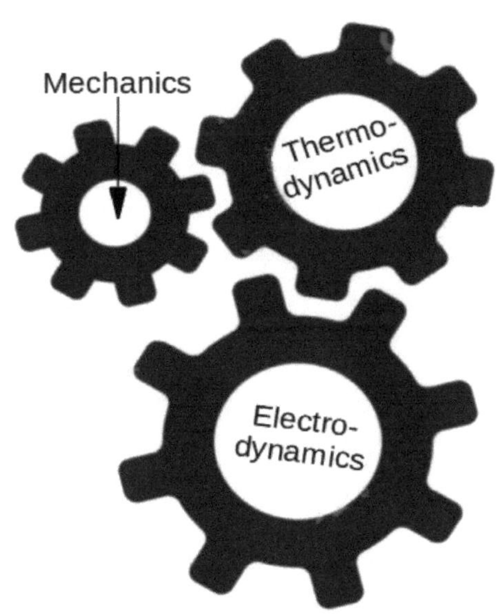

Printed and published by:
BoD – Books on Demand, Norderstedt

ISBN 9783755713753

Content

Preface for the Young Generation

Humans are not prisoners of fate, only prisoners of their own mind -
Franklin Roosevelt

Since the beginning of the information age, we can look back on a century of errors and excesses, but also of successes. We can see many things better today than we did decades ago, if we are only willing to undertake the effort of serious study. The problem remains of translating knowledge into charitable action.

In a time of climate change and the necessary energetic reorientation of our economy from fossil fuels to CO_2-free energy sources, we should know more about the functioning of the cosmos and the physical basis of its energy production. Theorists try to convince us that they already know everything, but that they cannot explain it to us properly. It would be too complicated. *"But as long as something is too complicated, you just haven't thought about it long enough"*, said Karl Popper. So we have to think again. Admittedly, this is difficult at a time like ours. Where can you find a place where you are undisturbed? To do this, you would first have to turn off the cell phone or, better still, look for a dead zone.

The word *cosmos* comes from the Greek and means order, and to establish an order obviously needs an organization. Since time immemorial, people have asked themselves: How did the world come about? Who did it all? In doing so, they reflect on their experiences from the environment and combine them in their imagination - often to create strange explanations. Surely there must be an organization or a creator behind it, just as they experience the organization of a social order themselves? But that would mean that there had to be a plan and a goal for the nature. Our knowledge of this is limited, our imagination limitless. But since ancient times we have been able to recognize neither purpose nor goal and this culminates in the sentence: *The ways of the Lord are unfathomable.* The problem lies in the complexity of nature, which far exceeds the complexity of our minds, in order to be able to recognize the operating principles in appropriate times.

The secret of the human mind, however, is its ability to abstract and classify in order to draw inductive inferences from it. Similar structures between the known and the unknown lead to the assumption that similar principles are hidden behind them, as with the known structures. In particular if these structures differ essentially only in terms of the scale of their dimensions.

Another special feature is that important insights leave a deeper mark on our mind than unimportant ones. This works in a similar way to an ant colony. All the ants whiz out to look for a source of food. Every ant leaves a pheromone trail on which it finds its way back. Over time, an intense pheromone trail forms where the ants were successful. In the case of ants, traces of pheromones evaporate over time if they are not constantly renewed. The traces in our mind also disappear again if they are not renewed. But we can store thoughts in libraries, that are important to us. This is what sets us apart from ants.

There is one more important difference. We don't know of any censors with the ants who erase or redirect traces on behalf of power. In order for people to become trapped in their minds, dictators control people's minds. With the help of ideologies or religions they set the limits of thought, prohibit thoughts going beyond that and claim absolute truth for themselves. At the same time they have a reward system with which they reward followers in order to secure their power. How this can be achieved, can be found for example in the encyclical **Pascendi Dominici gregis** by Pope Pius X from 1907. It was not only a guide for Catholicism, but also for communist dictatorships, although you will certainly not find a reference.

In order to channel people's thoughts, dictators need spiritual leaders who set the spiritual limits. From the 20th century we know a lot of such leading figures as Marx, Lenin, Mao, Einstein and others. Former GDR citizens know the West TV ban and Catholics know the *Index librorum prohibitorum.* Yes, science also knows such intellectual leaders as Albert Einstein or Stephen Hawking, whose importance is exaggerated by declaring them to be geniuses. Controlling this, is the job of Martin J. Rees, Live Peer in the *peer review system* and member of the **Pontifical Academy of Sciences** of the West**.**

The statements of such leading figures must not be questioned and with their statements you can stifle any counter-argument. „What Einstein says it has to be right, right? " And the opponent falls silent. Particularly in physics we know the peer review system, which, under the pretext of quality assurance in science, suppresses articles by researchers in renowned journals that are unpopular in this days, and gives the public the impression that science has been aligned. They speak of a mainstream and thus fakes a scientific consensus.

But since the introduction of the Internet, this has not been so easy. The number of dissident scientists in the field of physics is growing. Meanwhile, there are thousands worldwide who rebel against the closed world understanding controlled by the Catholic Church. It is not only the physical abuse of representatives of this power to deplore, but above all the spiritual abuse.

This mental abuse consists in selling fantasies as realities and generating miracles in order to gain power over people who believe in authority and thus to make them prisoners of their thoughts. This can go as far as attempting to enforce erroneous beliefs by force.
This can only be countered with a comprehensive general education, which, however, is still denied to many people for social reasons. You would think that nowadays, in the age of cell phones and the internet, everyone would have access to education. But it is not so easy with an oversupply of information, where truth and lies compete with each other.
It becomes particularly difficult when it comes to places and events that are almost never accessible and are therefore beyond direct observation. These places are ideal for false reports. Since time immemorial, this includes the macrocosm but also the microcosm, the world of atomic nuclei.

But since the middle of the 20th century this macrocosm has been moving more and more into people's consciousness, first as a playground for human fantasies and now more and more as an object of physical investigations, because we understand that we are part of this cosmos. In doing so, we come up against the abusively generated limits of an internally contradicting spirit, which can only be overcome in a democratic discourse. But these limits are usually not perceived in a static environment. I only became aware of these limits after the social change in my homeland.

Now the encyclical *Laudato si'* by Pope Francis in 2015 with *"the concern for the common house"* has just opened the door to a new understanding of world's nature. There it says under §79:

> *»In this universe made up of open systems that communicate with one another, we can discover myriad forms of relationship and participation. «*

Was the Pope aware of the revolutionary explosive power of this sentence when it was written? Because in doing so he opened the boundaries of the spirit as well, which for the Catholic Church had been firmly established for centuries. It reminds me of Günter Schabowki's press conference with the famous note in 1989 that went beyond the physical limits of my homeland.

In this book, I will be guided by that principle.

The cosmos is an open system and our knowledge gain is based on the inductive conclusion of the self-similarity of structures and their technical confirmation.

I consciously use the term *cosmos* instead of universe, since scientifically we can only survey a finite open part of this universe, which interacts with other parts that cannot be surveyed. With this I distance myself from the idea of a closed, expanding cosmos and at the same time also from speculations about non-verifiable cosmological questions about the beginning and the end, as well as not exactly definable terms.

In order for our knowledge of nature and the cosmos to develop, an understanding of democracy is also required in science, namely that in the end, despite all opposition, collective reason will prevail if our species is punished because people have a talent for reason, even if they often neglect this talent in times of abundance. Religions as well as today's understanding of theory are closed systems in which experiments are disqualified for their confirmation, the gain of knowledge needs an open system that is allowed to question theories. Science is not scholastic, it must be allowed to change. It only thrives in a democracy. If the 20th century is said to have been the century of reconciliation between faith and science, we can state that this led to an Ideologisation especially of theoretical physics. This Ideologisation went hand in hand with a growing arrogance and resistance to criticism on a part of its representatives.

This development is expressed in the term "dissident scientist". A list of alternative theories and criticisms of contemporary physics brings together around 10,000 scientists[1]), most of whom developed their ideas in isolation from one another. In real life, at the end of the 20th century, democracy triumphed over the ideological limits of dictatorship, but in intellectual life we are currently

---

1   Jean de Climont - *The Worldwide List of Alternative Theories and Critics;*
    https://books.google.de/books/about/The_Worldwide_List_of_Alternative_Theori.
    html?id=KnzBDjnGIgYC&redir_esc=y

observing the opposite tendencies in American Trumpism, but also in the increasing political radicalization in European countries. There intolerance is on the rise again worldwide.

It is time to reflect on the virtues of the 19th century Enlightenment, to combine them with the experiences of the 20th century, so that we can face the challenges of the 21st century.

I dedicate this essay to this issue as an offer for a better understanding of physics without unnecessary ballast from inconsistent theories that cannot be supported by verifiable facts. One of the fundamental paradigms of Theoretical Physics of the 20th century is the idea of the symmetry of the world. However, symmetry is at odds with dynamics.

All dynamics are movement and we learn the basic laws of movement in mechanics. Electrodynamics is high frequency mechanics and thermodynamics is low frequency mechanics. But only in thermodynamics do we learn that the world is not symmetrical and that the energy is distributed in it. If we want to explain the world, we have to understand that nature works differently than in the textbook with defined partial rules, but that all mechanisms of nature work simultaneously. The asymmetry of dynamics applies everywhere in the world, no matter how I try to explain it.

We will first deal with Ilya Prigogine's groundbreaking ideas and its consequences which raised thermodynamics to a new level of quality. In this light we will have to deal with the three of Albert

Einstein's most important works, which promoted the belief in symmetry, in order to arrive at a scale-invariant picture of electro-dynamics.

In doing so, I put electrodynamics and thermodynamics at the center of consideration, as the atomic structure of the world is based on the positive and negative charge of masses.

Jena anno 2021

1 The Flow of Energy in Open Dissipative Systems

»Everything that is intelligent has already been thought, you just have to try to think it again! « - Johann Wolfgang von Goethe

Physics remains largely incomprehensible to the layman because it is formulated in the language of mathematics. This is also effective for a long time, because facts can be represented more briefly by equations than if they had to be described verbatim. But the range of mathematical expressions is often insufficient to adequately describe a physical issue. In this way, idealized models of physical facts are made that can be calculated, but which therefore often deviate greatly from reality. The result is that with today's specialization in the individual disciplines, a belief in theory has spread among physicists that elementary fundamentals are often forgotten.

Academic physicists who are far from practice, believe that they have to invent new laws of nature instead of listening to the laws of nature and entering into a dialogue with it. In doing so, "*they let themselves be guided by the idea that the best theories are beautiful, natural and elegant,*" Sabine Hossenfelder says in her book *Das hässliche Universum*[2]). What beautiful is, must be true, is a widespread opinion among theorists. Beauty would distinguish successful theories from bad ones. That is a misbelief,

---

2   S. Hossenfelder – *Das hässliche Universum,*
     https://www.fischerverlage.de/buch/sabine-hossenfelder-das-haessliche-universum-9783103972467

truth is an evaluation, it cannot be calculated. How devoid of meaning such a belief is, is shown by the fact that the theories took on increasingly bizarre forms. Sabine Hossenfelder writes about it in her book:

>*»In my twenty years of studying theoretical physics, I have seen most of the scientists I know make careers studying things that no one has ever seen. They hatched insane theories like that our universe is only one in an infinite number of universes that together make up a "multiverse." They invented dozens of new particles and declared that we were projections of a space of higher dimensions created by wormholes that connected places far apart«*

But the physicists' ideal of beauty has fundamentally deviated from the artistic and aesthetic ideal of beauty since physics was over-mathematized at the beginning of the twentieth century. The symbol of mathematical beauty is the equation that expresses *symmetry,* while the ideal of beauty in the fine arts is expressed in the golden ratio and the Fibonacci spiral, in typical asymmetries. A hundred years ago, physicists took for granted that the fundamental laws of nature were mirror-symmetrical and that they would not distinguish between left and right.

Fig.1: The projection of the Fibonacci spiral and the golden ratio onto galaxy M51 in the constellation of Hound

They also believed that nature would strive for an equilibrium. This equilibrium is expressed in the mathematical equation too and where the symmetry was not so clearly recognizable, appropriate transformations were used and this was then called the theory of relativity, as if equality were the only mathematical relation. This mathematical poverty of physicists when describing dynamic processes is quite striking. It usually ends with point mechanics, but it has to be continued with thermodynamics and electrodynamics instead of viewing them as other disciplines.

A particularly striking example of this is Albert Einstein's handling of the dynamics of Maxwell's equations, which marks the transition from classical to modern physics. The culmination of this development should be the super symmetry of the Standard Model of particle physics. Eventually this theory was abandoned because none of its predictions were true.

Even in this days, many physicists interpret their subject area using equations as symmetries. The most significant example of this is Einstein's most famous formula, $E = m{\cdot}c^2$. The difference between the speed of light c and the speed v of a solid body is that c is not a vector. Light is distributed evenly throughout the room if no special technical measures are taken. So all directions are equal. A common light source is a *dissipative system*. You have to take this into account in the mathematical formulation.

So there is a misunderstanding that one can convert energy into mass and mass into energy. However, a mass is the carrier of the energy, from which follows: Without mass there is no energy. Hence Einstein's formula has to be written like this:

$$E \Rightarrow m \cdot c^2 \ \text{ or } \ E \Rightarrow p_e \cdot c$$

meaning that electromagnetic energy is distributed throughout the mass at the speed of light. There is a causal relationship with the movement of an electron and the triggering of an electromagnetic wave thru mass. But a wave is the result of the collective behavior of many particles of that mass. The impulse p_e is practically the force-free stimulating information that is distributed in the system with the constant speed c over the force field, which connects all atoms with one another, so that all particles behave in the same way. As a result of our analytical gaze, we have forgotten the cohesion of matter in structures. I use the term *stigmergy*[3]) for this behavior from biology. Only when the speed is braked at the phase boundary of the receiver does occur quasi the tsunami effect, the effect quantum of force. With this explanation, the paradox of the schizophrenic wave-particle dualism in quantum mechanics is resolved.

How this affects the energy at the receiver, describes

$$E \Rightarrow h \cdot v.$$

We call a dissipative form of energy *entropy* because it is distributed in all directions. The greatest effect is achieved by the explosion of an atomic bomb with its electromagnetic pulse, which triggers a thermodynamic storm. The energy carrier is the mass of the surrounding air molecules and atoms, which are ultimately responsible for the mechanical destructions. The speed of distribution of the electromagnetic pulse depends on the electromagnetic properties of the matter according to the relationship

$$c \Leftarrow \frac{1}{\sqrt{\varepsilon \cdot \mu}}$$

---

3   **Stigmergy** derives from the Greek words στίγμα stigma **meaning** "mark or sign" and ἔργον ergon **meaning** "work or action." It is a mechanism of indirect coordination, through the environment without any goal.

1 The Flow of Energy in Open Dissipative Systems

It is extremely remarkable that in modern physics the symmetry-breaking thermodynamics was completely disregarded in the mathematical handling of it. In the past, when they considered thermodynamics, they did so in a closed system, but never in an open system far from thermodynamic equilibrium.

Disregarding thermodynamics, modern physics has hit a dead end, as Lee Smolin noted in his 2007 book *The Trouble with Physics*.[4]) Not that there had been no attempts to get out of this impasse, but the mindsets of physicists were already hardened. Inspired by Ilya Prigogine's groundbreaking findings in the field of "nonlinear thermodynamics" and *dissipative structures*, for which he received the Nobel Prize in Chemistry in 1977, a few outsiders dealt with the phenomena of structure formation in Nature. In the nineties of the last century a couple of German books were published that described this item, such as *Grundprinzipien der Selbstorganisation*[5]) edited by Karl. W. Kratky in 1990 or *Chaos und Kosmos, Prinzipien der Evolution*[6]) by Werner Ebeling and Rainer Feistel in 1994. How much physicists are still attached to the belief in the divine symmetry of the world, we read there on page 31 under processes of structure formation:

»In terms of physics, our world was at the beginning of evolu-

---

4   L.Smolin - T*he Trouble with Physics;* https://www.amazon.de/Trouble-Physics-String-Theory-Science/dp/061891868X

5   K.W. Kratky - *Grundprinzipien der Selbstorganisation;* *https://www.amazon.de/Grundprinzipien-Selbstorganisation-Kratky/dp/ 3534109716*

6   W. Ebeling & R. Feistel - *Chaos und Kosmos, Prinzipien der Evolution;* https://www.amazon.de/Chaos-Kosmos-Prinzipien-Werner-Ebeling/dp/386025310 7

tion, with the so-called Big Bang, in a state of maximum symmetry. In the process of the evolution of the cosmos, the original symmetry was broken more and more, and new structures are constantly emerging. «

and a few sentences further we read the question:

»How did it happen that the complexity observed today developed from the initial chaos?«

From this we must draw the conclusion that the author defines chaos as the initial state endowed with the highest symmetry and he further concludes that chaos must have a creative potential in itself. From this conclusion he derives the theory of *self-organization*, i.e. the Deus ex Machina without energy. But here we want to question what this creative potency means physically. Even if the term self-organization is often found in connection with *dissipative structures*, I will differentiate myself from it. An organization is founded with a specific goal. With regard to an organized system, we differentiate between self-organization and *external organization*, whereby the goals can compete. You cannot speak of that at the particle level. The mass of the atoms is connected to one another by the force field of each individual particle, but nothing more.

Nowadays these topics are subsumed under the term *nonlinear physics*. This keyword is generally understood to mean research into dynamic processes with feedback, which is a bit misleading, because the terms electrodynamics and thermodynamics existed in classical physics before the symmetrization craze of modern physics broke out. Non-linearity is usually understood to mean equations of a higher order. The description of these phenomena by means of partial differential equation systems, which describe processes with feedback, to which Maxwell's equations belong, is more accurate. But here the development paths diverge.

While the development of modern physics began with its analytical separation of particles and symmetrization, the true character of Maxwell's equations lies in the description of electromagnetic energy dissipation via a system of force-coupled particles. But that was not recognized at the beginning of the 20th century. A real cultural struggle developed between the particularists and the continuouists, which, based on the misinterpreted ether experiment by Michelson and Morley, decided in favor of the particularists. [7])

---

7   J. DeMeo - *Dayton Miller ' s Ether-Drift Experiments : A Fresh Look;*
     https://www.semanticscholar.org/paper/Dayton-Miller-%E2%80%99-s-Ether-
     Drift-Experiments-%3A-A-Fresh-Demeo/
     101c2a023e2e9731212c00694568ff29b7968ed6

1.1 The Need for a New Understanding of Mathematics.

With this culture war in physics at the beginning of the last century, mathematics also got into a crisis. It lacked the appropriate tool to represent the discontinuity. The mathematical foundation of physics is the idea of infinitesimal calculus, namely the calculation of continuous curves by approximation using straight lines. These straight pieces are reduced infinitely until the length of the curve piece reaches a fixed limit value. Such curves are called *rectifiable*. When I was a student, we were trained in calculating integrals and differentials until we couldn't think of anything else. The problem now was to represent discontinuities and discrete quantities using continuous rectifiable curves. This should now be realized by the frequency analysis of wave functions, a bulky and unwieldy tool when it comes to discrete particles, especially because stable particles cannot be represented by superimposing waves.

Another problem arose. Not all natural curves are rectifiable. There are so-called *monster curves*, the approximation of which by straight lines does not converge to a limit value. We will come back to this later when analyzing Brownian motion.

How do you represent dynamics with the static means of equations? Of course, with coupled differential equations. With a system of coupled differential equations, there is no need for rectification. This means that breaks and discontinuities are tolerated. Instead, there is a demand for scale invariance. Scale invariance means that when the space cells are reduced in size, their properties do not change despite the breakpoints. We then speak of self-similarity. What scale invariance and self-similarity means can be studied by zooming in a Mandelbrot set. The Mandelbrot set has its origin in complex dynamics of a simple *algorithm.* This is the gateway to nature's fractal geometry. In general, algorithms are better suited to represent dynamics, since the behavior of coupled differential equations is often unpredictable, as the Conway game of life proves, which I reported in my book *Modern Astrophysics Meets Engineering*[8]). The Conway game of life is best suited for a simulation on a computer, in that the state of each room cell is calculated depending on its neighboring cells and is continuously displayed as a movie. By changing the rules of the game, you can model various dynamic processes. Of course, equations are the best way to describe symmetries. As a result, mathematicians with little physical expertise have imagined the world to be symmetrical, thus formalizing physics. But symmetry and dynamics are contradicting approaches. In the present essay, however, the main focus is on the self-similarity and scale invariance of nature. We will find that the methods of differential and integral calculus reach their limits when it comes to scale invariance.

The mainstream physicists have stuck to the concept of symmetry for over 100 years. Interesting, however, is a requirement that Lee Smolin made in his book *The Life of the Cosmos* as early as 1997. There he writes under the chapter *'Light and Life*[9]):

---

8   M. Hüfner – *Modern Astrophysics meets Engineering;*
     https://www.bod.de/buchshop/modern-astrophysics-meets-engineering-mathias-
     huefner-9783751920186
9   L. Smolin – *The Live of the Cosmos;* https://www.amazon.de/Life-Cosmos-Lee-
     Smolin/dp/0195126645 ; p.25 down

Physics, however, is in this days miles away from that, and he could not break free from the traditional thought structures that Einstein had given. They were carved into the textbooks as dogmas and so he remained a prisoner of these ideas.

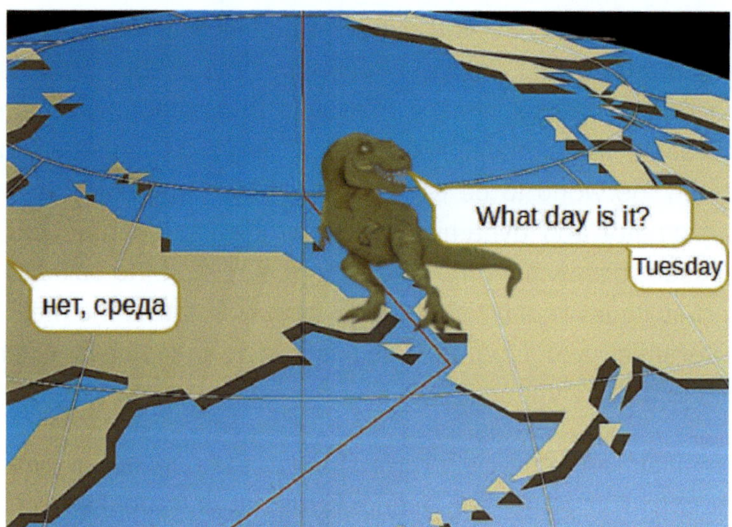

Fig.2: Relativity of time

Particularly blatant in modern physics is the way dimensions are dealt with, which reached its inglorious climax with string theory. Even in this days, relativists regard time as a dimension and thus as an independent quantity. Anyone who has been around the world today knows that time depends on place (Fig.2), but relativists insist that time is independent of place and that it has the meaning of a dimension.

According to the definition of space, however, this would mean that time must be perpendicular to the path. One look at an analog watch should be enough to convince anyone who doubts that time and path are interdependent, otherwise there would be no movement. In the big bang model it is even set absolutely. Of course, we engineers describe phenomena in space and time, but we call that a *process* and not four dimensional space-time. In a process, time remains a discrete quantity dependent on location. We express this fact through a description of the state that is linked to a scene and a concrete time interval.

Mathematics has developed tremendously in the last hundred years, but modern physics has preserved the level of mathematics from a hundred years ago. Computer algorithms are suitable for the representation of dynamic processes, especially when it comes to feedback processes.

For the description of dynamics, algorithms are much better suited than coupled partial differential equations.

The famous mathematician Alan Turing paved the way for the new way of thinking in 1937 by using a theoretical machine to grasp the concept of *predictability*. Today each of us has a practical machine with which we can (provided we have the necessary qualifications) model dynamic processes on the basis of two-valued logic. Antony Peratt[10]) succeeded in simulating cosmic dynamics for the first time in 1986 when he simulated the behavior of two ion currents (*Birkeland currents*) on the computer and as a result he obtained a vortex that took on the shape of a spiral galaxy. (Fig.20 p.96)

1.2 The Central Meaning of the Asymmetry of Thermodynamics

Our most important physical knowledge so far is that Einstein's energy relation is to be considered under the aspect of dynamics and that the light propagation has no preferred direction, which has the consequence that the light is distributed in the room.
Now we have learned that in a closed framework the second law of thermodynamics holds that the change in entropy is positive. Thermodynamics is, on the one hand, an expression of molecular mechanics and, on the other hand, an expression of molecular electrodynamics. In other words:

Thermodynamics is the central bridge element between mechanics and electrodynamics, and our world is consequently asymmetrical in its dynamics.

But how can a dynamic be maintained in a closed system? The equipotential bonding is completed within a finite time and then the system is dead. The dynamics of a flow require an inflow and an outflow within a given framework. I renounce the set-theoreti-

---

10   A. L. Peratt - *Evolution of the Plasma Universe: II. The Formation of Systems of Galaxies;* IEEE TRANSACTIONS ON PLASMA SCIENCE, VOL. PS-14, NO. 6, DECEMBER 1986;
    https://www.academia.edu/9156671/Evolution_of_the_Plasma_Universe_II_The_
    Formation_of_Systems_of_Galaxies

cal definition of an open system and content myself with its intu-
itive meaning.

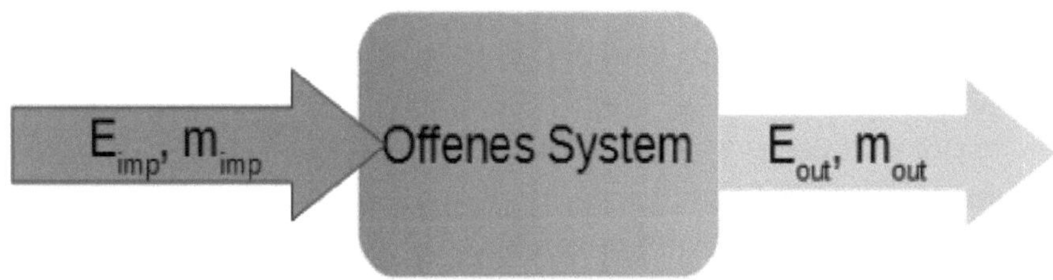

Fig.3: There are no absolutely closed systems

Let's start by describing an *open system*. It has an entrance and an exit for the exchange of energy and mass with its surroundings. Every real system is more or less open. A closed system is an idealization. Its borders must be absolutely impermeable.

The simplest of these open systems is a frequency converter. It receives electromagnetic radiation at an input frequency and emits it again as electromagnetic radiation at an output frequency because in the steady state case it consumes energy for its own movement and radiation.

Each atom unit represents a frequency converter, regardless of its internal structure and charge. If a layer of atoms is illuminated by light, part of the light is reflected from its frequency spectrum, the rest is absorbed, converted into mechanical energy of charge carrier and emitted as thermal radiation. We notice a warming of the atomic layer. An atom absorbs energy from the nano wave range and starts to vibrate in resonance. These resonances then emit radiation in the microwave range again. Provided that the energy is retained, follows according to Planck and Helmholtz's theorem of energy conservation

$$h_1 \cdot v_1 = h_2 \cdot v_2.$$

So if a short-wave energy radiates in and a long-wave energy comes out. The effect on the atom must have increased and it must have been set in motion. We measure a temperature increase at the point of irradiation. There are obviously many different quanta and not just Planck's constant for an electron. To imagine an effective quantum, think of a hammer drill, or how a bridge vibrates under the lock step of a battalion marching over it.

Now Ilya Prigogine (Илья Романович Пригожин) has made a few basic statements about the thermal behavior of open systems. In an ideally closed system, the entropy increases as the usable energy decreases. The second law of thermodynamics says.

$$dS/dt \geq 0$$

or to put it casually: In a closed system, disorder can only increase. The typical example is the compulsive hoarder, a person who dumps the garbage in his apartment.
  In the 1970s, Ilya Prigogine expanded equilibrium thermodynamics to include the thermodynamics of open systems far from thermodynamic equilibrium.

In an open system there is no longer any static equilibrium, but there is a steady state, in which the energy introduced can be equal to the energy emitted and the mass introduced can also be equal to the mass emitted.
What can we say about the entropy in such a system? The change in entropy of an open system is the sum of the entered

change in entropy and the internal change in entropy minus the released entropy. We combine the difference between the entered change in entropy and the change in entropy given to form the external change in entropy.

Then we can write:

$$dS_{System} = dS_{inp} + dS_{int} - dS_{out} = dS_{ext} + dS_{int}$$

For this Ilya Prigogine () received the Nobel Prize for Chemistry in 1977 [11]), because he was able to explain why open systems can organize themselves, because although $dS_{int} \geq 0$, dS_{ext} is arbitrary. We can distinguish between three cases:

1 **The order in the system is established.**
   More entropy is removed from the system than is generated internally.
   $dS_{ext} < 0$, $|dS_{ext}| > dS_{int} \rightarrow dS_{system} < 0$

2 **A steady state has been reached.**
   $dS_{ext} < 0$, $|dS_{ext}| = dS_{int} \rightarrow dS_{system} = 0$

3 **The order in the system falls apart.**
   There are two reasons: an internal and an external.

   a) Less entropy is removed from the system than is generated internally.
   $dS_{ext} < 0$, $|dS_{ext}| < dS_{int} \rightarrow dS_{system} > 0$

---

11   I. Prigogine – *Time, Structure and Fluctuations;* Nobel Lecture, 8 December, 1977 ; https://www.nobelprize.org/uploads/2018/06/prigogine-lecture.pdf

b) The external entropy is greater than the internal.
$$dS_{ext} > dS_{int} \rightarrow dS_{system} > 0$$

This *order of nature* is not a new discovery, it was just reformulated by Prigogine. We already encounters its philosophical formulation in the ancient Indian Vedas[12]), it was adopted into the Samsara (Cycle of life) of Hinduism and anchored in the cosmic Trimurti principle, symbolized by the gods Brahman as creator, Vishnu as preserver and Shiva as destroyer. In Christianity only the trinity of a catholic-patriarchal monotheism remains. However, this was robbed of its original symbolic meaning and reinterpreted in the sense of absolutist claims to power. An expression of a changed understanding of nature is also the encyclical *Laudato si'* by Pope Francis, because he too had to recognize that climate change belongs to category 3a of Prigogine's thermodynamic law, the internal reason of destroying.

I am not a believer, but I pay respect to such an ancient and so deep wisdom, even if it comes in the guise of a religion, because it seems that the majority of people only have respect for things that have been declared sacred.

It is therefore my concern to give thermodynamics the central place it deserves in my further remarks. But first let's take a look at the ideas that have troubled physics, because the academic physicists since Einstein, put the human mind with its projective point of view above natures reality.

---

12 Veda = Knowledge, ancient Indian oral traditions originated from 3000 years ago and were written in Sanskrit from the 5th century onwards.

2 The Original Sin of Physics

»Imagination is more important than knowledge because knowledge is limited.« - Alber Einstein

Mechanics has been thought symmetrically since I. Newton, L. Euler, J.L. Langrange, M. Hamilton, H. Poincaré, A. Einstein and P. Dirac. These are ingrained thought patterns. Mechanics is the first thing taught in physics education. It is the mechanics of a mass point, the movement of which is completely independent of its surroundings. None of these theorists has ever thought that point mechanics could be integrated into a larger theoretical building of fluid dynamics with inter-dependencies between flow cells.

The 19th century was marked by enormous experimental advances in the field of electrodynamics, which M. Faraday, J.C. Maxwell, W. Weber, H. Helmholtz, W. Thomson [Kelvin] and H.A. Lorentz, with the exception of Weber, thought of as fluid dynamics.

In 1911, the first Solvay conference on the subject of *Theory of Radiation and Quantum* ended a real cultural battle between a fluid electrical worldview and a mechanistic atomistic worldview, represented by theorists such as Poincaré, Planck and Einstein, where theoreticians were victorious. Three works on problems in physics by the then 26-year-old patent engineer Albert Einstein, which were regarded as revolutionary at the time, also contributed significantly to this, with which he supported the ideas of the editor of the *Annalen der Physik*, who was none other than the well-known and influential Max Planck.

Albert Einstein consolidated the mechanistic-atomic worldview with his essay *Zur Theorie der Brownschen Bewegung,* by tracing the chaotic trembling movements of small pollen in a liquid back to the molecular vibrations. In particular with his work *Über einen die Erzeugung und Verwandlung des Lichtes betreffenden heuristischen Gesichtspunkt* Einstein supported Planck's idea of the quantum of action (Planck's constant) as a natural constant and thus the development of quantum mechanics. In return, Planck contributed to the dissemination of the theory of relativity that Einstein had developed in his essay *Zur Elektrodynamik bewegter Körper.*

As a result, under the influence of the newly founded Papal Academy of Sciences with its aim of reconciling science with faith, a development began in the coming decades that made mathematics from a tool to the dominant element in physics. Theoretical physics decoupled more and more from the natural sciences and the traditional idea of symmetry in closed systems became more and more decisive compared to natural scientific knowledge.

2.1 Analysis of Einstein's Work under the Premise of Energy Dissipation in Open Systems

We want to take up the three themes from 1905 on the character of light, Brownian movement and the dynamics of solid bodies under the premise of open energy-flowing systems. We are aware that today, after more than a hundred years, we have a much greater wealth of knowledge available on the Internet than Einstein had in his days through libraries in his area. As a result, it is not just a matter of evaluating his personal academic performance, but also always evaluating the social context in which his ideas stood. We cannot reconstruct which knowledge was personally available to him and when and how he assessed it. Judging from his resume, he had fundamental problems with authority, was a spirit of contradiction, and with little social skills. But he had apparently a suggestive persuasive power.

2.2 About Light and Mass

His work on a heuristic point of view concerning the generation and transformation of light earned him the Nobel Prize, although his explanation of the photoelectric effect is contestable. To illustrate Einstein's argument a little, here are a few quotes from the essay.

> *»There is a profound formal difference between the theoretical ideas that physicists have formed about gases and ponderable bodies and Maxwell's theory of electromagnetic processes in an empty space.«*

Einstein used the culture war between an analytical and a synthetic approach at the beginning of the 20th century. The difference in the way of viewing lies essentially in the scaling of the viewing frame. Maxwell's theory never started, as Einstein claims here, from an empty space; he always assumed the ether as the medium of propagation. It is an allegation on the part of Einstein. All atoms have neighbors and are connected to one another via a force coupling. Newton already saw the force coupling of two masses over great distances. Since the force coupling of several bodies was mathematically not mastered, later mechanics simply ignored it. Maxwell, on the other hand, relied on Descartes' vortex theory, which Helmholtz later expanded and thus influenced many scientists.

We read on by Einstein just there:

> *»The energy of a ponderable body cannot break down into any number of any small parts, while the energy from a point-like light source emitted light beam according to Maxwell's theory is continuously distributed over an ever-growing volume. «*

The energy generally is bound to a ponderable carrier. However, when Einstein speaks of the energy of a light source, he is referring to its intensity. He also speaks of a point-like light source that emits a light beam, although a transmitter must have an adequate expansion for its wavelength and no light beams can be seen in the cosmos, which soon astonished Wilhelm Olbert. (Olbert's Paradox) What we call light beam is the shortest distance between the receiver and the light transmitter. Einstein now draws the conclusion from the above sentence:

> *»It now seems to me… that the energy of light is distributed discontinuously in space. … it consists of a finite number of energy quanta localized in spatial points, which move and without dividing, can be absorbed or generated as a whole.«*

Since he sees the room as empty, he lacks the transmission points for the transmission of impulses between the light transmitter and the light receiver. So he resorted to Newton's light particles. Max Planck was able to verify the equation $E = h{\cdot}v$ on his receiver, a spectrograph, and Planck's constant is generally thought to be a natural constant on which quantum mechanics is based. However, his equation is not restricted to atomic scales.

Nowadays every do-it-yourselfer has a hammer drill in the house that works on the principle of impulse transmission, the quantum of which can be easily calculated if you know the impact frequency. Einstein would then come along with a sandblasting fan in order to stay in this picture. Every do-it-yourselfer also observes that the impact of the hammer drill only becomes apparent when it is placed on the hard wall. This is analogous to the photoelectric effect. This also only occurs at the phase transition to a denser medium as a result of the change in speed of the wave front of the light wave because force is the product of atom mass at the wave front and the change in its speed.

There is no need to introduce photons as particles to physics at all.

By equating the energy of the transmitter to the energy of the receiver, you can even determine the supposed mass of the light particles, which is formally mathematically possible, but from a physical point of view is utter nonsense. But the *big bang model* assumes that the mass emerged from the light. Here, according to *fiat lux*, cause and effect have been swapped.

The concept of *mass* is used in particle physics in a misleading way. The fact that Einstein speaks of heavy and inert mass in his *equivalence principle,* shows that for a long time it was not possible to clearly distinguish between mass and *force*. Inertia and heaviness have something to do with forces that work in the gravitational field. Masses, on the other hand, are uncountable quantities. You can buy a commodity by unit of measure or unit of mass. Quantities are counted and masses are weighed using beam scales, for which a calibration mass is required in order to exclude the influence of gravitational force. Forces are generated by *charges*, but charge-neutral masses do not exist in nature. This is due to the asymmetry of the positive atomic nucleus and its negative, movable electron shell. Because we can count these on the level of elementary particles, the concept of mass becomes superfluous on the level of atoms.

The decisive property of the elementary particles is their charge and only then their number. For the emission of electromagnetic waves, it is their negative acceleration, because as long as electrons, for example, rotate freely in the atomic shell, no electromagnetic waves are emitted. The larger the oscillating charge units, the larger the emitted wavelength.

> The playing of an organ with its organ pipes of different sizes leaves a special sensual impression. The simple rule here is that halving the pipe length results in a tone that is twice as high, i.e. a tone with twice the frequency. This means that the pipe that sounds an octave higher than the low C is only half its length, i.e. 4 '(= 1.20 m); the pipe of the next octave 2 '(= 60 cm), and so on. In fact, the pipes are a little longer than the specified number of feet: In addition to the length of the swinging column of air, there is the tapering pipe foot at the bottom and a reserve to be flexible in the mood.

The relationship between mass and force has a long history. Thales of Miletus first discovered in amber around 600 BC that surfaces polarized by friction exert attractive forces on one another. Since amber is called *Ελεκτρον (Elektron)* in Greek, the effect of attraction has been referred to as electricity since William Gilbert around 1600. If you have to use forces for the division of material bodies, you can assume that opposing forces are responsible for the cohesion of bodies. Forces can only be differentiated according to strength and direction. Nevertheless in textbooks, they are still differentiated into four types according to their origin. The forces that keep the celestial bodies on their orbit are called gravitational forces, although Coulomb's law of electrostatics and Newton's law of celestial mechanics differ only in one proportionality factor. In the 20th century, through the division of physics into different chairs, academic principalities seem to have developed, which keep their borders closed and defend their territories by all means.

However, if gravitational forces and electrical forces are measured with the same measuring principle, namely with the Cavendish torsion balance, there is only a difference in the amount of these forces, but not in their origin. As early as 1836, Fabrizio Mossotti described gravity as a residual force of electrical force[13]. Although

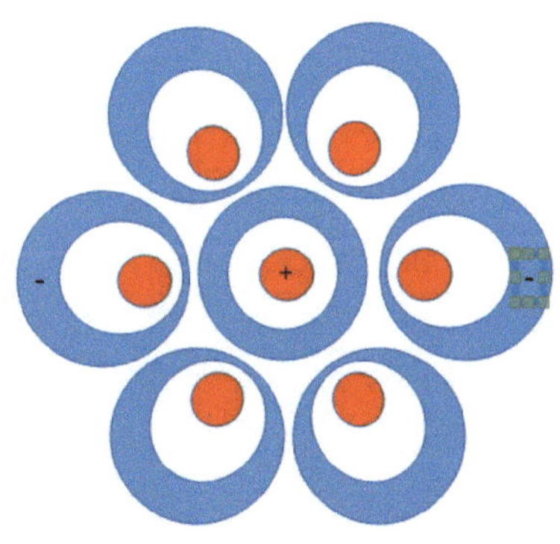

Fig.4: The electrical cause of gravity

both forces have the same origin, Albert Einstein separated gravitation from electrical force in his paper on *The Electrodynamics of Moving Bodies*. In these days it should be clear to everyone that the forces between the atomic nucleus and the atomic shell are forces between positive and negative charges. The atomic shell and the atomic nucleus seem to form a neutral body. But the charges can be shifted within an external force field. This means that atoms of the same type, but also molecules, tend to join together to form larger, massive bodies (aerosols), as Fig.4 shows.

We want to designate the resulting charges as *bound charges*, because they can only move within the atom. In contrast to this,

---

13   F. Zöllner - „*L'attraction universelle elle mê peut découler comme une déduction des principes qui règlent les forces électriques.*" Leipzig 1882 Commisionsverlag S. XXVI

there are still the *free charges* on the surface of a homogeneous solid body, in which charged ions or free electrons can accumulate there.

The forces resulting from free charges can be shielded by opposing charges in contrast to the small forces of bound charges. We use the term massive or mass here because we have no information about the number of atoms and molecules that combine to form *aerosols* or solids as a result of stigmergy and their resulting charges. So we have two different types of charge. On the one hand we have the charge bound in the atomic nucleus, which makes atoms appear as dipoles in the external field and on the other hand we have free charges on the surface of solids or in liquids and gases when atoms are ionized, i.e. electrons are removed from the atomic shell. The removal and addition of surface charge naturally affects the energy content of the aerosols.

We speak of aerosols for particle sizes from nanometers to micrometers, i.e. particle diameters that are two to six powers of ten above the atomic diameter. As a rule, these are particles in the order of magnitude of hundreds of thousands to trillions of atoms in a lattice. These associations usually have a negatively charged surface, which allows them to float in the negatively charged earth field and at the same time form condensation nuclei for clouds. Unfortunately, little is known about the effects of masses of this magnitude.

Now we can also answer the question of whether light can be deflected by gravitational forces. As we have noticed, light is not a massive particle but an impulse that propagates force-free in a homogeneous medium at constant speed. As long as the speed is not slowed down by a compression of the propagation medium, there is no possibility of deflecting the light, since it has no charge that another charge could act on by means of a force.

Even in this days, the representatives of modern physics believe that symmetry, expressed by reversible uniqueness, is the main property of matter, that energy and mass can be converted into one another due to the Einstein's energy-mass relation $E \Rightarrow m \cdot c^2$ and that mass would grow continuously with its increasing speed. In fact, it can be shown that if you formally replace the mass by p / v in the above relation and do a few mathematical operations, you get the form

$$E \Rightarrow \frac{E_0}{\sqrt{1 - \dfrac{v^2}{c^2}}} \quad \text{and then from this follows} \quad m \Rightarrow \frac{m_0}{\sqrt{1 - \dfrac{v^2}{c^2}}} .$$

But there is a problem. What does m = p / v refer to. Is this relationship even permissible? It comes from the exact form $\vec{p} \Rightarrow m \cdot \vec{v}$. but a vector division is not defined at all and so the derivation of the above relations is a **faux pas** as well as using time as an independent vector with the fatal consequences.

Because the momentum is a conserved quantity, a force is used to separate a mass into at least two bodies, e.g. into an electron and a proton (rocket principle). If the first body remains at rest, this does not mean that the mass of the second accelerated body increases. This is what the equivalence principle says, because the inertial mass is equal to the heavy mass. If you differentiate between heavy and inert mass, it has nothing to do with the mass itself, but with the forces that act on the massive body. This means that the greater the moving apparent mass of a body, the greater the lost energy, according to the directional relation that is distributed to the environment and not that a charged body moving at almost the speed of light would get additional mass.

May be one Electron would become two or more?   No, the cause is the resistance that the electromagnetic field between the atoms opposes to the movement of the charged body, because Wilhelm Weber discovered the connection between the speed of light and the electromagnetic properties of matter as early as the 19th century.

When a mass of atoms is accelerated, then innumerable charges always move with it. These charges induce a magnetic field that opposes the movement with a drag force. This mechanism is registered as a virtual increase in mass, although the mass of the atoms remains constant. With a careful calculation, Paul Marmet has proven that the virtual increase in mass is actually due to the magnetic field generated when charges are moved.[14])

2.3 The Brownian Motion

The Einstein's second essay in 1905 *On the motion of particles suspended in still liquids required by the molecular kinetic theory of heat,* which caused a sensation at the time, was the one on *Brownian motion.* Einstein described it as a result of the internal energy of molecules, which up to now is generally accepted be-

---

14   P. Marmet - *Fundamental Nature of Relativistic Mass and Magnetic Fields.*
      https://www.newtonphysics.on.ca/magnetic/index.html

cause it seemed to explain some of the properties of these movements. But this explanation does not stand up to critical scrutiny.

We have above reported that Einstein's most famous equation $E = m \cdot c^2$ from a thermodynamic point of view is not an equation in the mathematical sense, but a directed relation $\Rightarrow$ that describes the distribution of electromagnetic energy to the surrounded mass at the speed of light. With this in mind, let us now consider the Brownian motion of microscopic particles in liquids and gases.

The Scottish botanist Robert Brown was the first to describe the strange movements of pollen in water when he looked at them under a microscope in 1827.

He attributed it to the life force (internal kinetic energy) of these pollen grains. It was not until 1905 that Albert Einstein took up this phenomenon again. You can imagine that Einstein also considered the phenomenon under equilibrium conditions here, in accordance with the view of the time. According to him, this movement is also erroneously referred to as Brownian molecular movement, although the mass of the water molecules compared to the observed pollen grains is about 20 orders of magnitude smaller. In this days we know that a water molecule weighs just under 3×10^{-23} g and a pollen grain is in the order of 3×10^{-3} g. If you compare a pollen grain with nowadays largest container ship, the difference in mass at least 14 orders of magnitude. So you would have to compare a million of these 400m long container ships with a grain of pollen. A million of these container ships in a line would circle the equator 100 times in a row.

Einstein's explanation, which is very unsatisfactory from today's point of view, is that the displacements of particles visible in the microscope are caused by the fact that the molecules constantly hit the particles from all directions in large numbers due to the osmotic pressure and their disordered heat movement, and in doing so, times the one direction, sometimes the other direction is more important. That should make even the greatest Einstein admirer wonder. Larger structures must be responsible for this.

The formula derived from Einstein using infinitesimal calculus simulates an accuracy that is not inherent in it, since the movement of the grains does not follow a rectifiable movement curve. Here failed the conventional mathematics, as it has designed for Newton's mechanics. The motion curve is broken up into a multitude of irregular pieces. Benoît Mandelbrot introduced the term *fractal* for such curves and thus opened up the subject area of fractal geometry. We will deal with this in Chapter 3.

But back to Brownian motion: It would depend on the diffusion constant and thus essentially on the temperature of the surrounding liquid. Only the changes in movement are not constant. Since the temperature in the macro range is in equilibrium with the environment and also shows no difference in the micro range

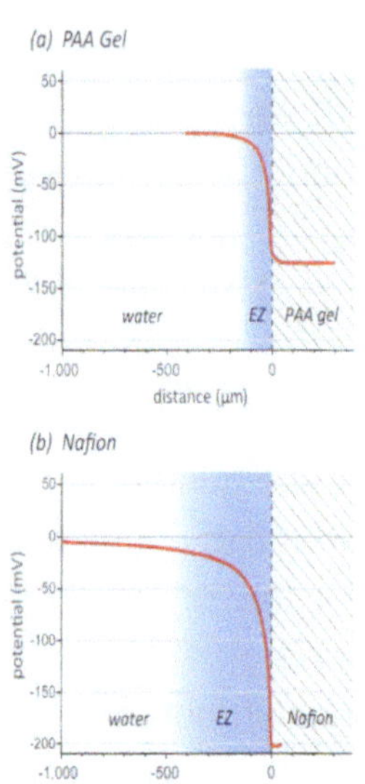

Fig.5: Electrical potential in the exclusion zone - reference: G. Pollack

around a pollen grain, the movement remains a mystery, especially since some properties remain unexplained. Strangely enough, this explanation is still accepted by science to these days.

However, Gerald Pollack[15]), one of the most important researchers in the field of water, found a completely different explanation for the Brownian movement. If you look into a glass of water, water looks the same everywhere and yet the structure of water molecules in the vicinity of the surfaces of other substances changes fundamentally.

When he experimented with micro spheres in water in the nineties of the last century and observed their distribution, after some time he noticed the formation of zones from which the micro spheres withdrew at interfaces with solid bodies. He called these zones *exclusion zones*, short EZ. His years of studies have brought the certainty that the exclusion zones or restricted zones observed could not be explained trivially. Measurements near gels in water showed that the exclusion zone was negatively charged. Consequently, there must be positively charged areas outside of this exclusion zone. Invisible currents seem to move through the water.

On hot summer days we expect a good thunderstorm with lightning and thunder towards evening when the clouds have piled up. Every child is taught that lightning strikes are dangerous

---

15   G. Pollack – *The Fourth Phase of Water: Beyond Solid, Liquid, and Vapor;*
      https://www.amazon.de/Fourth-Phase-Water-Beyond-Liquid/dp/096268953X

electrical discharges between the clouds and the earth, but also between the clouds themselves.

Hence, like batteries, clouds need to store electricity. Now everyone knows that a car battery needs acid. The main difference between a car battery and a cloud is the size, which affects the charge density. But the fact that there is no charge compensation even in a water glass may surprise you.

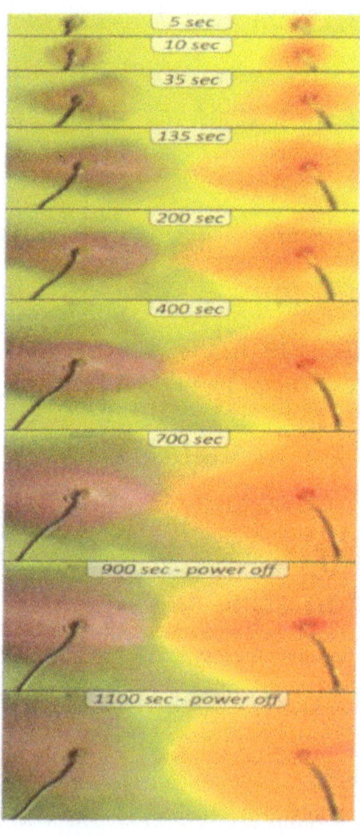

Fig.6: Charge distribution in water made visible with pH-sensitive dye - reference: G. Pollack

If we put two electrodes in water, we can split the water molecule into a positive H ion and a negative OH group. That should have stuck from chemistry class. I experimented a lot with electricity as a boy. For this I used 4.5V flat batteries and checked their voltage with my tongue for lack of a voltmeter. If the battery was still usable, the electrodes tasted really sour. The H ions are responsible for the sour taste. Well, there are better acid sensors than the taste test. The litmus tincture has proven its worth in addition. Depending on the pH value (the pondus hydrogenii), the tincture changes color. The pH value can then be read off from a calibrated color scale. Here, acidic water tends to be red and alkaline water tends to be blue. The degree of acidity is now dependent on the number of free protons occurring in the water and the basic character of the water is determined by the proportion of free, negatively charged molecular groups. If you put such dyes in water, you can clarify the micro-fluid properties of the water.

Fig.6 shows the mode of action of pH-sensitive dyes, the time course of the charge distribution in water between two current-carrying electrodes, on the right the cathode around which the protons, indicated by the red color, collect, and on the left the anode. Within 11.6 minutes, a positive and a negative field developed around the electrodes, which is retained even after the current has been switched off.

Einstein's formula for Brownian motion is based on a state of equilibrium and the trembling of the molecules is due to an internal free energy. As long as the temperature remains constant, the system should neither receive nor lose energy. However, a glass of warm water should give off energy to the environment and a glass of cold water should absorb energy. If the glass is in the room for a while, the temperature should be balanced, according to the classical theory. But as long as the sun is in the sky, water is constantly absorbing electromagnetic energy from the environment. Absorbed light energy brings an aqueous system out of balance in terms of charge. We observe this in plants, for example. How else is a chemical reaction

$$6\,CO_2 + 12\,H_2O + light \rightarrow C_6H_{12}O_6 + 6\,O_2 + 6\,H_2O$$

like photosynthesis to be understood? So when water is out of balance, a theory that relies on equilibrium has to be rejected.

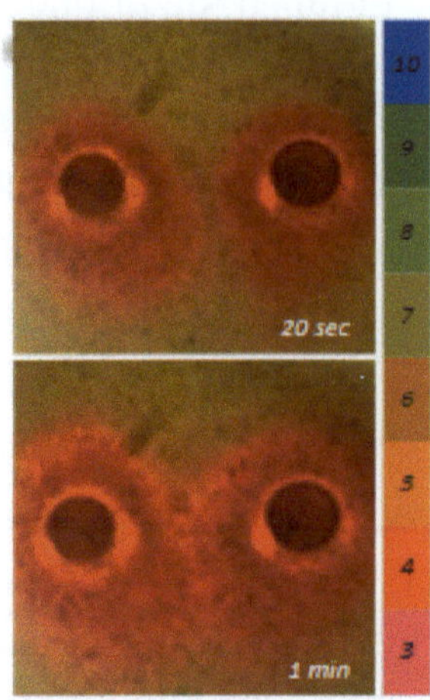

Fig.7: Two micro-spheres with their exclusion zone - reference G.Pollack

In countless experiments, Pollack and his team have found that water molecules at the phase boundaries with hydrophilic substances form layers of *poly-water* under the influence of light. The hydrogen forms bridges to the oxygen atoms so that an oxygen atom is always surrounded by three hydrogen atoms. This creates a liquid crystal with a hexagonal structure, which is also found in snow crystals. The water there only has the possibility of shifting as one layer of poly-water against another. During this formation process, excess protons and foreign bodies are shifted outwards, as can be seen in Fig.7. The pH changes from 5 to 4 and we get a yard of EZ water around the micro spheres.

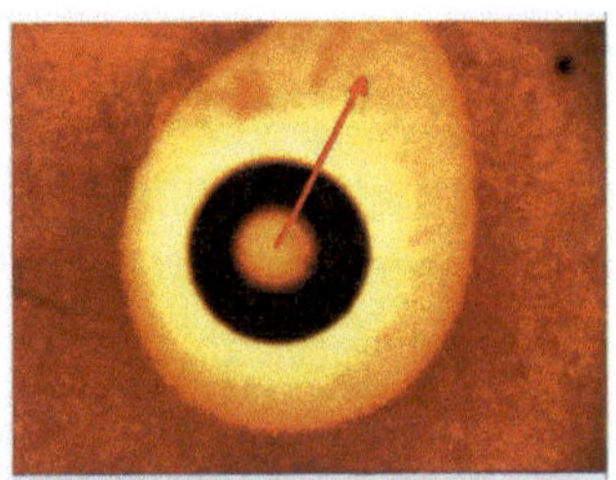

Fig.8: Micro-sphere with exclusion zone for EZ water, light comes from the top right and the movement goes in this direction - reference G. Pollack

These exclusion zones have two basic properties: charge separation and order of the molecules. Entropy is released into the environment. The superfluous protons are removed from the exclusion zone. The charge density in front of the exclusion zone increases, which can be seen from the more intense red color. Charges partially generate forces among one another. In this way, other pollen grains with their restricted zones are kept at a distance.

If the energy flows evenly onto a pollen grain from all sides, the exclusion zone should be concentric. In practice, however, this cannot be achieved, as can be seen after a minute. So there will always be a resulting direction of force and the micro-sphere will move accordingly, which of course will also trigger movement in the neighborhood. If the incidence of light changes from diffuse to directional, as shown in Fig.8 on a micro-sphere with a diameter of 0.5mm with its exclusion zone with additional incidence of light from the top right, the exclusion zone on the side of the light incidence widens significantly and superfluous protons are ejected. As a result, the micro sphere and the exclusion zone start moving in the direction of higher positive charge density.

If several particles are involved, the movement becomes complex, as the exclusion zones influence each other and the movement constantly changes the regional charge distribution. So it is not the individual water molecules that cause the movement, but the effects of the restricted zones around the pollen grains that prevent the pollen grains from coming too close. What can you learn from it now? If we look at the chemical reaction of photosynthesis and the behavior of water in the vicinity of a wetted surface, we can see that we have an open aqueous system, into whose entrance energy flows in the form of electromagnetic radiation, which creates an order in the system with the release of entropy builds up.

This is precisely the thermodynamics in open systems that Ilya Prigogine has already described, which leads to dissipative structures, and Gerald Pollack was actually able to prove that water in the exclusion zone on a phase border forms a liquid crystal with entropy delivery.

Although the young Einstein also dealt intensively with thermodynamics, and as early as 1903 published an essay entitled *A Theory of the Fundamentals of Thermodynamics*[16]), he did not get beyond the derivation of the second law of thermodynamics in equilibrium systems, which is perfectly understandable for a young man, if you does not see him as the forward-thinking genius how to present him in public.

---

16 A. Einstein – *Eine Theorie der Grundlagen der Thermodynamik;*
      https://onlinelibrary.wiley.com/doi/abs/10.1002/andp.19033160510

2.4 Einstein's Dynamic of Movable Bodies

The title of his most spectacular paper from 1905 should actually be *Electrodynamics of Solid Bodies*, because electrodynamics is the study of the influence of electrical forces on the motion processes of bodies. It would probably not occur to anyone that there is also a dynamic of resting bodies.

If he had complied with his father's wish and studied electrical engineering in order to be able to join the company *Elektrotechnische Fabrik J. Einstein & Cie* instead of applying to the federal polytechnic school in Zurich, he would be in his most important essays ten years later likely to come to very different conclusions. You could understand this work as an opposition to his father, because it shows the complete lack of his understanding of the dynamic. The first sentence already proves this:

»It is well known that Maxwell's electrodynamics - as it is currently understood - when applied to moving bodies leads to asymmetries which do not seem to adhere to the phenomena.«

The fact that this work nevertheless caused a sensation is due to the intellectual environment into which it fell. Regarding this intellectual situation in his environment, it should be noted that on the one hand Arthur Schopenhauer had a strong influence on the young Einstein. With his book *Die Welt als Wille und Vorstellung,* in which he criticized Immanuel Kant's book *Kritik der reinen Vernunft,* he shaped the idea of relativity as a symmetrical relationship between observer and object, which Einstein transferred to movement.

On the other hand, Auguste Comte's demand that sensory perceptions should be the starting point for thinking and philosophizing, was popular at the time. This demand, together with George Berkeley's subjective idealism, created a peculiar philosophical climate in which both a theory of relativity, could survive, characterized by the identity of image and object as well a quantum mechanics that negated causality in the micro-range and elevated coincidence to the rank of objective reality.

In the spiritual power center of the Catholic Church at the beginning of the 20th century, took place a confrontation between the reactionary forces and the modernists, which was expressed in the encyclical *Pascendi Dominici gregis* of 1907. Subsequently, however, under the influence of George Lemaître, in the Vatican efforts were made to reconcile the science with faith. So Albert Einstein became a sorcerer's apprentice on the leading strings of a Catholic Church that was regaining its strength after a period of Enlightenment in the 19th century, in which George Lemaître got the leading position in the scientific academy.

> *»We want a fireworks theory of evolution. The last two thousand million years are slow evolution: they are the smoke and ashes of bright but very rapid fireworks.«*

> *»There is no conflict between science and religion.«*
> – Georges Lemaître
> New York Times, February 19, 1933

The centuries-old dispute as to whether the earth or the sun is in the center of the world was settled with Einstein's theory of relativity. The censorship of the peer review system of a scholastically frozen sect of scientists, led by the papal academy of science, is still at work on physics in this days.

But what is it that makes this Relativity so fascinating? It is the unusual, the fourth dimension, that arouses spiritual fantasies. This is the bridge that scientists have to cross in order to fulfill Lemaître's concerns. In the Middle Ages it was the Immaculate Conception of the Virgin Mary. In quantum physics it is the dualism between wave and particle. If science should to serve as the handmaid of religion, faith must at least control reason, whether it is spiritual or esoteric is secondary. Fellowship is crucial.

As early as 1878, the Leipzig astrophysicist Carl Friedrich Zöllner caused a sensation when he supported the American Henry Slade with spiritual and esoteric reflections on the fourth dimension. For example, he argued like this: *In order to get a mirror-symmetrical figure from a two-dimensional figure on a sheet of paper, I have to fold the paper through a third dimension. Consequently, I need a fourth dimension in order to get a mirror-symmetrical body from a three-dimensional body.*

But that did not earn him any recognition among the enlightened scientific community. On the contrary, that seriously damaged his reputation.

A generation later, the scientific community did exactly the opposite and swallowed this toad. When the already well known physicist Max Planck, chairman of the respected German physical society and later member of Papal Academy of Sciences too, disseminated Albert Einstein's ideas, this spiritual idea was accepted much more readily. The reason was probably that this time the idea was mathematically better packaged in that it came along as a Lorentz transformation and in general not very much was known about the mathematical group of transformations. While the Euclidean transformation changes the point of view, there are transformations that depict reality by reducing the dimensions by one.

We are familiar with the perspective depiction of an object with its distortions, which, however, correspond exactly to our viewing habits. The Lorentz transformation is however a depiction with distortions that do not correspond to our world of experience and which we therefore perceive as unusual. As long as you don't take the picture for reality itself, that's fine too.

The matter only becomes a fall into sin through the positivist attitude of selling images as realities to the ignorant public. Another famous example that characterizes this esoteric attitude towards quantum mechanics is the thought experiment with Schrödinger's cat, which crouches as a zombie in the box and only gets the concrete state of living or dead when the box is opened. The quantum mechanics around Max Planck also gained increasing influence in the papal academy onwards from 1927.

When Einstein turned against quantum mechanics with the words *God does not roll the dice*, his career was over. The same thing happened to Stephen Hawking in 2014 when he admitted the non-existence of black holes after years of wrestling with Leonard Susskind. His admission remained largely unknown to the public and, unimpressed, a Nobel Prize was awarded in 2020 for the "discovery of a black hole" in the Milky Way. After we have discussed the physical basics, we will deal with the inner workings of galaxies in Section 4.6.

2.5 The Business with the Unusual

The externally determined belief usually controls people's lives, the more the lower their level of education, the more so. This belief does not have to be anchored in religion at all. I believed in science for many years before I started to question my area of expertise myself. It was the internal contradictions that puzzled me. From then on I began to use my intellect without outside guidance, as Immanuel Kant would put it.

As long as you need someone else's management, you remain in a state of immaturity. Roosevelt's quotation on the limits of the mind therefore aimed at this immaturity of the believer. What does the underage believe most? - In fairy tales and miracles. Especially in times of need, people were conditioned by sacred books and urged by priests to believe in miracles.

Miracle workers seemed to be able to work these miracles by virtue of their overwhelming spiritual power. Healing miracles in particular took up a great deal of space. A distinction was made between good (church-friendly) and bad miracle workers. The former were idolized and the latter demonized. I remember the burning of witches in the Middle Ages as a particularly bad derailment of human belief, where women with medical knowledge were persecuted and demonized.

The Catholic Church still has a custom to canonize people who are said to have performed good miracles. In 2005 Pope John Paul was canonized. The miraculous belief that united them all under the roof of the Church was the faith in the *Immaculate Conception of the Virgin Mary.* There are repeated reports in the press of child prodigies whose IQ is compared to Einstein and Hawking, whose legacy is the belief in the "black hole" that warps space-time, and that doesn't even require the Catholic faith. But what benefit society does have from such behavior? The same that society has when children are told about Santa Claus or the Easter Bunny. Its to hold parts of society in immaturity.

It is therefore not surprising that people's interest is generally geared if not to look for miracles but at least for the extraordinary, in order to either delight in it or to work out a special position as a person of "great spiritual strength". Einstein is committed to the latter.

> *»When we're working on something, we get off our high logical horse and sniff the ground. Then we cover our tracks again in order to increase the likeness to God.«*   — Alber Einstein

So if science and belief are to come closer to one another, as Lemaître had intended, the Church must create such human idols that occupy a special position and proclaim something unusual and that must be done in a language that most people do not understand so that the Status of being chosen is underlined. Mathematics is ideal for this, although it does not always have to be used exactly. Unusual can be moved into areas that cannot be recorded with our measuring devices. Where do you find better conditions for this than in the subatomic realm and in the cosmos?

That is exactly the work field of modern physics, the search for the extraordinary in the inexplicable. How do you want to find something extraordinary that cannot be grasped with sensors? It has to be invented. Once you have an idea, you develop it into a theory and then have it confirmed by an experiment that is as complex as possible. The problem, however, is the interpretation of the experiment.
Then the contradictions to the theory usually become visible. So it has not yet been possible to prove the theory of relativity unequivocally through an experiment.   But from this theory Karl Schwarzschild and Stephen Hawking derived the black holes, huge invisible gravitational monsters that occupy people's fantasies like no other idea. The intensely glowing galaxy cores are still being painted over by black circles, although Hawking announced in 2014 after years of conflict with Susskind, which went down in physics history under the name of the "*Black Hole War*", that there is no event horizon and with it there is no black hole behind it either.

It is like the proofs of God. Intellectual products remain intellectual and their fate is tied to the authority of their inventors. If they were to materialize, that would be magic, which is why I like to call such people magicians. But this magic obviously has great entertainment value and is therefore an economic factor. A new type of entertainment has emerged from it, the science fiction film. Movies such as *The Silent Star, Planet of the Apes* or *Star*

Wars are movies that have shaped millions and millions people's ideas of the cosmos, although these movies only transport current usual human problems into an unusual world.

But that theme falls into the field of spirituality and esotericism, while a natural science always starts from the material appearances and makes a theory of them. Herein lies the fall into sin of physics, that its representatives, following the example of Einstein, in the arms of Catholic spirituality, have said goodbye to the goals of a natural science of working out an objective worldview. Nowadays they understand science not as knowledge but as $cience. But if you only think of fame and private profit, you forget the real concern.

We find the extraordinary in myths and legends, not in science.

3 Dynamic Structures and Stigmergy

Structures are characterized by the arrangement and connections between their elementary parts. We call this loose association stigmergy.

While the theories of modern physics are maximally suitable for the entertainment industry, the achievements of our technology are mainly based on the physical findings of the 19th century. So let's first forget all the theories of modern physics that were invented since Einstein and ask what dynamic structures we encounter in nature. πάντα ῥεῖ says Heraclitus of Ephesus "Everything flows" and everything turns, symbolized by the eight-limbed Dharma wheel of Buddhism. This knowledge of over two thousand years is still valid today.

Rotation and *translation* determine all dynamic structures in the world regardless of how we look at it. The translation can be seen as an excerpt from a rotation with a comparatively very large radius. A dynamic structure always arises from the collective interaction of individual parts, which we can currently grasp with some reliability on scales from 10^{22} to 10^{-12} meters, from the galaxy to the elementary particle. There is currently only one conclusive explanation for the collective interaction between atoms and suns. It is the interacting forces between the structure-forming members. This creates an indirect coordination of the members by their neighbors. We call this behavior *stigmergy*. The term comes from the Greek words stigma (στιγμα) for marking and ergon (εργον) for work.

We only find organization at a higher level in the development of dynamic systems, who set themselves apart from the goals of an external organization with their own goals of self-organization. Organization requires a certain amount of awareness in order to distinguish between one's own and those of others goals.

Nature would not be nature if, in addition to cohesion, it also had breaks in its phase transitions between different material properties and states of aggregation. But infinitesimal calculus fails at phase transitions.

Once a principle has been successful, nature repeatedly applies it to all scales. So we find self-similar fractionally broken structures everywhere. Our thoughts also follow the principle of self-similarity when we use the inductive inference to generate knowledge. At this point I would like to remind you of the great mathematician Benoît Mandelbrot, whose ideas had a lasting impact on mathematics in the 20th century. So far, his ideas have not yet found their way into physics, but they will certainly gain even greater importance in the coming years in connection with stigmergy. Nature is governed by the principles of fractal geometry, a simple example of which is presented here.

3.1 Ideas from Fractal Geometry

The infinitesimal calculus is based on the idea that a smooth and steady curve can be approximated by increasingly smaller line segments and that the limit value of this approximation tends towards a fixed limit value what we call convergence. All of physics uses the tool of infinitesimal calculus and ultimately maps its laws in polynomials and converging mathematical power series in order to be able to represent them on the computer.

However, we have seen with Brownian motion that the infinitesimal calculus fails for the first time in a calculation because the motions do not produce regular smooth lines, as we are used to from Euclidean geometry. On closer inspection, nature rarely offers smooth lines and surfaces that can be described by continuously differentiable functions.

In reality, the surface of all real bodies is rough due to the granular structure of matter and macroscopically visible structures are often repeated on lower scales. We want to explore the consequences of this by examining the length of the coastline of an island. It is an example of the measurement problem that geophysicists have.

It is perfectly clear that the length of a line is the distance between a start and an end point on a straight line. Because a coastline, like a Brownian motion line too, is not a straight, but a very irregular line, we have to approximate it using straight measuring lines. Usually the integral calculus tells us in theory: the orbital integral as a limit is then the length of the coastline or motion line. In practice, however, the length of a coastline depends on the choice of straight measuring line segments. Because of its roughness, a coastline does not approach a limit value, but we find similar structures again on a smaller scale, which increase our measurement result. The reason is that we did not take into account the scale of our consideration when calculating.

As soon as we zoom into the image of our island, new, similar structures become visible that were previously hidden from us.

We find that the measurement result is quite arbitrary and the coastline is not rectifiable. Lewis Fray Richardson found a two-constant relationship for many coastlines in 1961[17]).

$$L(\Delta s) \sim \lambda \cdot \Delta s^{1-D}$$

To understand the meaning of this relationship, we have to look at a self-similar, nowhere differentiable, artificial curve that the Swedish mathematician *Helge von Koch* presented as early as 1904. The Koch curve represents a fractal, the length of which can be determined directly. This curve is described by iteration as follows:

> First, a segment with a given length is given (generator). Then remove the middle third of the line and build an equilateral triangle over it with the length of the removed line (iterator) which is one third of the original length. This increases the curve by four thirds. In the next step, the iterator is applied to each of the four resulting route sections. This iteration is now repeated as often as required. Then Δs continues to shorten, but the number continues to increase and with it its total length. This results in two parameters $N = 4$ and $\varepsilon = 3$.

In the case of the Koch curve, the length of the curve can be calculated using Richardson's relationship.

$$L(\Delta s) = \Delta s^{1-D} \text{ with the factor } D = -\frac{\log N}{\log \varepsilon}$$

---

17   L. F. Richardson – *The Problem of Contiguity: an appendix of statistics of deadly quarrels* in General Systems Yearbooks 6 pp. 139-197

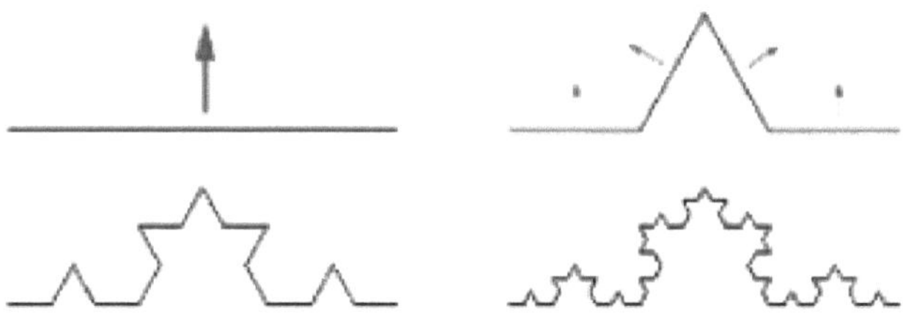

Fig.9: Koch curve

If the length of the original segment connecting the two ends of the curve is *a* and after the iteration step *n* the curve length has increased by the factor $\left(\dfrac{4}{3}\right)^n$ the length of the curve is

$a\cdot\left(\dfrac{4}{3}\right)^n$ and for $n \to \infty$ follows $\lim\limits_{n \to \infty}\left(\dfrac{4}{3}\right)^n \to \infty$

Now we connect three segments of length *a* with each other to form an isosceles triangle and apply the Koch algorithm to all three sides. If we now ask about the area of the Koch snowflake, we immediately see that it cannot grow beyond all limits.

The area is obtained by adding three times the area below the Koch curve to the area of the equilateral triangle. The exact value is of no interest here, as the calculation can easily be found under the keyword *Koch's snowflake*. It is the ratio of circumference to enclosed area or surface to volume, which is important here for self-similar fractals in relation to free surface charge to bound charge. With the knowledge of chapter 2.2 we can now conclude:

Due to its fractal nature, the free surface charge can vastly exceed the volume dominated gravity.

This knowledge is of fundamental importance because it explains why the electrical properties of the plasma are predominant over large stretches of the cosmos and why there can't be gravitational monsters like black holes.

It is about the same effect why clouds weighing tons remain in the sky and only rain down often after they have discharged by lightnings. It explains dust explosions too and also makes star novas in the cosmos more understandable.

3.2 Aerosols as Base Structures

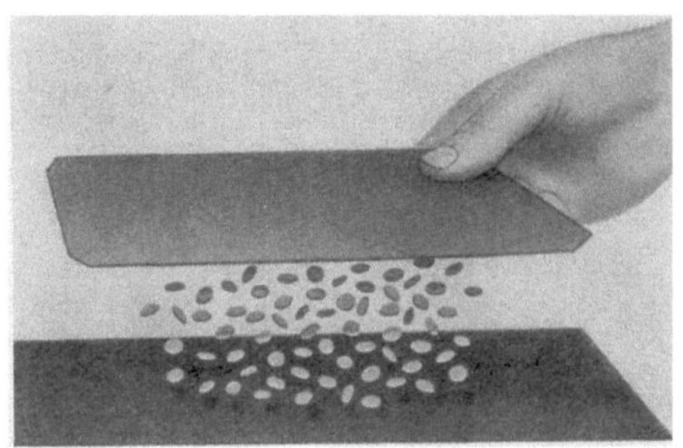

Fig.10: Confetti dance in the electric field

We denote as aerosol a suspension of fine solid particles or liquid droplets in air or another gas. Aerosol grains (solid or liquid) can be understood as self-similar fractals with a small volume but a large surface. The consequence is that the free electricity on the surface far exceeds the bound charge in the form of gravity. So aerosols are electrostatic charged. That is the reason why they do not sink to the ground in the earth's field, but float in the air. As a result of their self-similarity, their dimensions can vary by orders of magnitude. The field strength surrounding them is decisive for their behavior. So we can also see the cosmic dust as an aerosol.

At this point I remember an experiment from my childhood that I was able to carry out with my electrical construction kit. To do this, we sprinkle the confetti snippets from the file punch onto the table top and rub a decelith plate with a silk cloth and then hold it over the snippets. Between the tabletop and the decelith board, the confetti snippets begin a lively dance. Fig.10 is reminiscent of Brownian motion in a water bath. This movement seems to be completely chaotic.

But as soon as energy flows through an aerosol swarm, these movements are rearranged according to Prigogine's theorem, if the change in entropy of the open overall system remains below zero. We observe this behavior with the smoke of a cigar, with steam over a teacup or the formation of clouds in the sky. Everywhere we encounter fractal structures in the dynamics of an aerosol swarm.

What we see rising as white vapor, are droplets with a diameter of at least 1 µm so that they can be seen at all. If the radius of a water molecule is 0.3 nm, there is room for around 40 billion water molecules in that mist droplet and space for around 8 million free charge carriers on its surface. However, since Coulomb's electrical force is theoretically 2.4×10^{39} times stronger than Newton's gravity, in principle a free charge carrier on the mist droplet is sufficient for it to rise from the equally charged tea cup. More about the behavior of water vapor can be found in G. Pollack's book *The Fourth Phase of Water*[18])

Fig.11: Vapor over tee cup
Photo: G. Pollack

---

18 G. Pollack - *The Fourth Phase of Water, beyond Solid, Liquid and Vapor;*
    https://www.amazon.de/Fourth-Phase-Water-Beyond-Liquid/dp/0962689548

3.3 Vortices as Basic Elements of Fluid Dynamics

As a boy I was always fascinated by the smoke rings that my cigar smoking grandfather blew into the air from time to time for my entertainment and I asked: *"Why do they stay stable over a certain period of time?"* He couldn't answer the question for me.

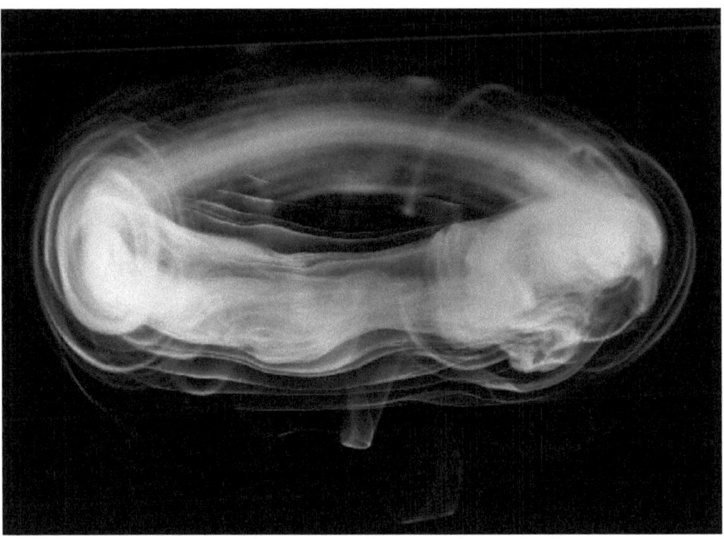

Fig.12: Smoke ring - Photo: Tobias Tschepe

Now I have a term for it -- *stigmergy.*
In fluid dynamics, this property is characterized by the Reynols number. At a critical speed, a laminar flow changes into a swirled flow because the forces increase perpendicular to the direction of flow. High flow velocities always lead to turbulence.

If we look at the smoke ring, we can describe it as a donuts-shaped torus, which arises from two mutually perpendicular movements that can easily be transformed into a plane. This movement can then be represented as a function of the radius of the torus.

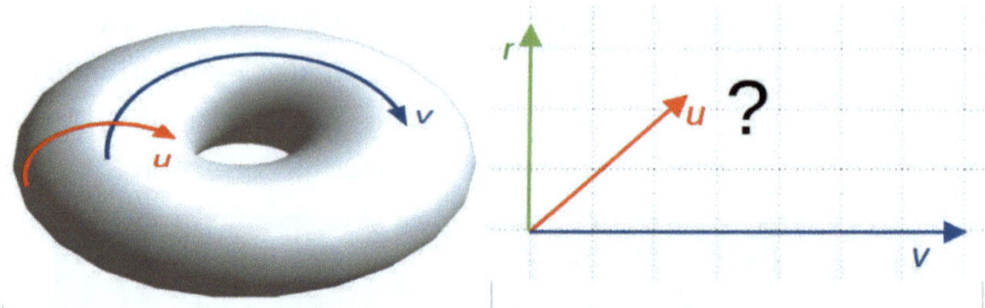

Fig.13: Torus-Transformation

Not only aerodynamic currents, but also electrical currents that flow with the velocity *v*, experience a force that is described in the mechanical flow theory by the *viscosity* and which is called the *Lorentz force* in electrical current circles, which deflects every current perpendicularly with the velocity *u* and gives the movement of every free stream a helical shape. Complex numbers can be used to mathematically describe this movement. In order to make it easier to understand, we will refrain from doing this here.

When aerosol particles are set in motion by external forces, this happens in an orderly manner, thanks to the stigmergic forces that exist between the individual particles and the release of entropy to the environment. Viscosity can explain mechanically stable vortices. But that item falls into the field of thermodynamics of open systems. The fractal nature of the material phases not only leads to differences in the potential of electrical charges between them, but also to different accelerations within aerosol currents, which in turn are described mechanically. There is special literature for this.

Now we cut open a vortex ring and consider its cross section, the actual vortex. The forms of appearance of the vortices are very diverse, depending on whether they are stationary $v = 0$ or whether they are moving and changing their shape. Even so, there are two basic types for vector u.

First there is the *irrotational vortex*, in the center of which the angular velocity of the flow $\vec{\Omega}=(0,0,ar^{-2})$ is one of the highest with $\vec{r}=(x,y,0)$.

The particle velocity $\vec{u}=\vec{\Omega}\times\vec{r}=(-a\,yr^{-2},a\,xr^{-2},0)$ is inversely proportional to the distance r from the axis. The particle speed is inversely proportional to the distance r from the axis. Then a particle in this vortex would not spin on itself. It would maintain the same orientation as it circles around the vortex axis. In this case the angular velocity $\vec{\omega}=curl\,\vec{u}=0$ is zero at any point off this axis. This means that the particles themselves do not rotate.

In relation to the vortex center, their rotational speed is the opposite of the vortex rotational speed.

In such vortices the stigmergic forces between the particles are small. An observer sitting on a particle observes that the vortex center revolves around the observer. The center itself remains free of particles. An example of an irrotational vortex is the tornado, which presents itself as a swirl of clouds. The eye of the tornado is clear of clouds.

The counterpart to this is the *rigid body vortex*, in which the individual particles all rotate around their axis at the same angular speed $\vec{\Omega}=(0,0,\Omega)$. If the vortex rotates like a rigid body – i.e. the angular velocity Ω is uniform, so that it increases proportionally to the distance r from the axis – a particle carried by the flow would also rotate around its center as if it were part of this rigid body. In such a flow the vorticity $\vec{\omega}=rot\,\vec{u}=2\vec{\Omega}$ is the same everywhere: its direction is parallel to the axis of rotation and its magnitude is equal to twice the uniform angular velocity of vector Ω of the fluid around the center of rotation.

An observer sitting on a particle always looks at a fixed vortex center. This requires a strong cohesion between the particles, because the greatest circumferential speeds are at the edge of the vortex, where the centrifugal forces are greatest. A vortex that rotates in the same way as a rigid body – cannot exist indefinitely in that state except through the application of some extra force, that is not generated by the fluid motion itself. So this vortex has been called *forced vortex* too.

Wilhelm von Helmholtz, who published his vortex theory soon in 1858, pointed out that as early as 1755 Leonhard Euler[19]) recognized that magnetic vortices behave like rigid body vortices.

A mixed form is the *Rankine vortex*, which corresponds to a rigid body vortex in the interior and which approaches a irrotational vortex in the outer area.

If we have always looked at the center of rotation from above, we have neglected the third dimension. A vortex has a vertebral axis around which it forms a vortex filament. In his mathematically founded vortex theory from 1858, Hermann von Helmholtz coined a remarkable sentence about this vortex filament.

Vortex filaments must run back into themselves or end at a phase boundary.

19   Histoire de l'Acad. des Sciences de Berlin. An 1755 p.202.

But that says that every vortex that does not meet a phase boundary forms a vortex ring like these smoke rings of my grandpa.

In the next chapters we will come across these vortex filaments and vortex rings as self-similar fractals on all scales.

4 The Helical Structure of a Vortex Filament

It is a mistake to believe that movement is an independent reality. There is always a mass what moves in time. Consequently, energy cannot be converted into mass because mass is the carrier of energy either.

If we look at dynamics in open systems, then we look at a rather small, approximately straight-line arc piece from a comparatively large vortex ring, say a circuit. The circuit is so large in relation to the induced magnetic vortex ring that we can usually neglect the curvature at the point of observation. We assume a potential difference between the input and output of the system so that dynamics can even get going. The potential difference creates a force that accelerates charges. Charges are usually carried by protons and electrons, the quantities of which are innumerable. An accelerated charge creates a second force that is perpendicular to the direction of motion, which we call the *Lorentz force*. A third force acts on parallel currents that compress them. They are called *z-pinch force*. When masses move through such a field, they do so with minimal energy. However, minimal energy means that the movement is powerless at a constant speed. In the force field described above, a charged body must then move on a helical path.

> This phenomenon was first observed by Kristian Birkeland at the Polar lights at the end of the 19th century and confirmed by measurements with space probes in the sixties of the 20th century, which is why these currents were named Birkeland currents in his honor.

Our analytical thinking has taken apart these two components of motion (rotation and translation) for mechanical understanding. In electrodynamics, these two components are brought together again. If you look at the projection of this movement in the x, y-plane, you get a wave movement from the screw movement. The basic theory of the wave behavior of matter was developed by Louis-Victor de Broglie in his dissertation in 1924, for which he received the 1929 Nobel Prize in Physics.

Just obviously forgotten that the wave movement in a projection plane is in truth a screw movement in space. Two years later, Erwin Schrödinger developed from de Broglie's idea the wave equation for quantum mechanics. The projection of the movement perpendicular to the vertebral axis provides the image of rotation as Descartes and Newton described it for the first time.

It is astonishing how the mechanical and the electrical perspective came into competition with each other and, at the latest, when it became clear that the sun and the planetary system is moving its orbit in our home galaxy, that should have led to a rethink.

Maxwell's theory came under pressure at the beginning of the 20th century when it became clear that electrons did not radiate energy under all circumstances when moving, namely when they circled the atomic nucleus on force-free orbits and gave off energy only in certain amounts when they switched to an energetically lower orbit. That was the reason to develop quantum theory. When the first satellites were launched into Earth orbit from 1957, however, they circled a center of force with just as little force. But it was too late to see a similar effect in this. While Maxwell's equations in the classical imagination say that any movement of charge creates electromagnetic radiation, obviously that only applies to negative acceleration. This becomes very obvious with the brems-strahlung of electron beams.

Maxwell's equations describe the energy distribution of an oscillating circuit in space. They describe this as an interlinked self-similar force-free vortex structure of a transmitter. But they describe neither the receiver nor the flow of electricity from massive bodies. In this respect, the desire to symmetrize this system of equations of the transmitter by means of a projective transformation, as represented by the Lorentz transformation, testifies to a physical incomprehension that can hardly be surpassed. Every dynamic is asymmetrical.

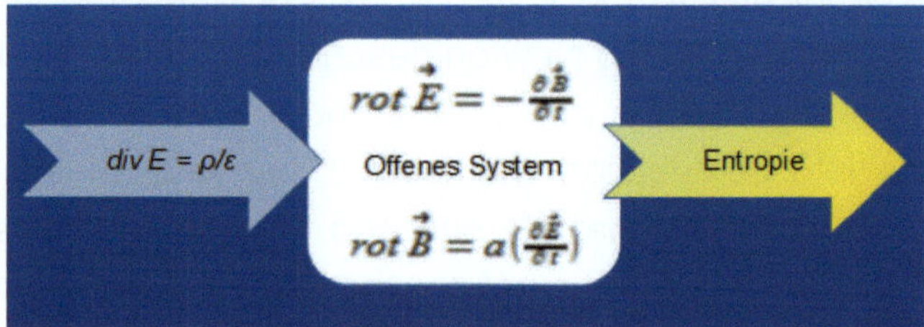

Fig.14: Representation of Maxwell's equations as an open system

While Einstein's relationship between energy and mass describes the general distribution of energy, Maxwell described the manner in which it is distributed. This takes place through coupled vortex rings. However, he did not specify any restrictions for the applicable frequency range. As a result, his equations apply from the shortest waves, the Gamma rays through visible light to thermal radiation and radio waves, in short over the entire electromagnetic spectrum. The field of thermal radiation and of course also statistical mechanics is covered by the entire thermodynamics. In this respect, the dispute over the predominance of the mechanical properties of the cosmos is only a Lilliputian dispute[20]).

4.1 About Newton's Dynamics of Massive Particles in the Cosmos

By massive particle flow we mean particles that contain protons as opposed to a pure flow of electrons in a wire. In general, we consider weighable masses to be electrically neutral because we refer to the earth potential, which we define as zero potential. But that's not true. The earth potential is negative and the change in field strength is around 140V/m near the ground. This tension results from gravity and the free surface charges. Surface charges are fed from the natural radioactive decay of the elements. The postive ionosphere, which is fed by the solar wind in 2000 km high, reaches into 60 km above the earth's surface. When considering the mechanics of masses, we have to be aware of the relationship between the bound charges and the free surface charges. I remind you that in the chapter on fractal geometry we learned that surfaces can be very large in relation to volume.

Because our world consists of negatively and positively charged elementary particles, we understand gravity in the tradition of the enlightened 19th century as an electrostatic force that our earth exerts with its gravitational acceleration on another bound charge. In the case of bound charges, the forces of attraction always predominate, since the charge carriers always form dipoles. We describe these dynamics with the means of mechan-

---

20   The Lilliputians fighting the Blefuskans because of disputes about which side they are to break the egg. From Jonathan Swift's novel Gulliver's Travels

ics, neglecting the charges. Free charge carriers behave differently and there we use electrodynamics. We distinguish the electrodynamics of the electrons in a solid body, where we can neglect the weighable masses, from the electroplating or the dynamics of a plasma, where we have to take the weighable masses into account.

Have you never wondered why Newton's apple falls back to earth but the moon stays in the sky? Well, our rockets are given the appropriate starting impulse so that they can reach orbit. But where does the starting impulse come from for all the heavenly bodies that are circling around each other? That is supposed to be done by the one-dimensional gravity of an Isaac Newton?

Nevertheless, it was a great achievement for that time to find the force F_r that a large mass M exerts on a small mass m at a distance r, based on Kepler's laws of motion of the planets, which are based on Tycho Brahe's observations. At a great distance applies

$$F_r \propto \frac{M \cdot m}{r^2}$$

(1)

We use the relation sign proportional ($\square \propto \square$) here instead of the equal sign so that we don't always have to write the constants. However, the above relationship, empirically derived in the observation framework, has a huge flaw in order to be able to apply as a general law. It has a singularity at $r = 0$, which would mean that if you imagine the mass M to be concentrated in one point, as is usual in point mechanics, the potential would grow beyond all limits, which would be equivalent to a hypothetical black hole. There is no such thing in nature. Real mass has everywhere a volume. You cannot concentrate all mass in one point, so you have to take mathematical precautions so that masses do not crash into each other at the speed of light.

First we have to put Newton's law in cylindrical coordinates. Then we move the large mass M according to its inertia to the origin ($z = 0$; $r = 0$) and m circles at a distance $\vec{r}$ around M. After Einstein we call this an *inertial system*. Then in this coordinate system an attractive force directed towards the center is negative and a repulsive force is positive. We must therefore write Newton's law in vector notation as a two-dimensional relation as follows:

$$\vec{F}_r \propto -\frac{M \cdot m}{r^2}\,\vec{e}_r \qquad (2)$$

where e_r is the unit vector of the radius. It should be noted that Newton's law is an idealization of the true relationships. Since the sun radius is very small compared to the distances between the planets (*sun radius to distance from earth to sun: 0.46%*), both masses can be understood as point masses. However, slight deviations have already been observed for Mercury and Venus. For Mercury the ratio of the solar radius to the distance of Mercury is already 1.2%. Therefore Wilhelm Weber made a correction to Newton's relation and added a limiter:

$$\vec{F}_r \propto -\frac{M \cdot m}{r^2}\left(1 - \frac{v_r^{\,2}}{c^2}\right)\vec{e} \qquad (3)$$

This is to prevent the two point charges from generating a force that grows beyond measure when approaching them at will. Finally he added a term, the 2nd derivative of the radius with respect to time, i.e. a radial acceleration b_r.

$$\vec{F}_r \propto -\frac{M \cdot m}{r^2}\left(1 - \frac{v_r^2}{c^2} + \frac{2r}{c^2} \cdot b_r\right)\vec{e}_r \qquad (4)$$

The radial acceleration determines the eccentricity of the ellipse and is responsible for the rotation of the perihelion. In these two extensions Wilhelm Weber saw the basis for the fact that the law of gravitation of all weighable bodies results as a consequence of the basic electrical law and he emphasized the importance of confirming this assumption for the whole of physics.[21]

Felix Tisserand[22], a French astronomer had found deviations in the perihelion of Mercury and Venus that could explain this factor. He stated a deviation of δ = +13.65" per century and for Venus δ = +2.86". The current forecast of perihelion rotation for Mercury is δ = +42". It was not until 1898 that Paul Gerber[23] succeeded in completely deriving the formula for the perihelion rotation. Gerber's formula for the rotation of the perihelion was formally identical to the equation later set up by Einstein. Paul Marmet confirmed Gerber's derivation without relativity theory only with the theorem of the conservation of mass and energy, which played no role in Einstein's principle of relativity.[24]

A riddle arises from the assumption that the Newtonian relation is supposed to describe an irrotational vortex from which the planetary system is supposed to have originated. If we want to as-

21 W. Weber - *Elektrodynamische Maassbestimmung insbesondere über den Zusammenhang des elektrodynamischen Grundgesetzes mit dem Gravitationsgesetz* in Webers Werke Bd. IV S. 481

22 F. Tisserand - *Sur le mouvement des planètes autour du Soleil d'après la loi électrodynamique* de Weber. *Compt. rend. 1872. Sept. 30.*

23 P. Gerber - *Die räumliche und zeitliche Ausbreitung der Gravitation.* in Stargard in Pommern, 1898. http://bourabai.narod.ru/articles/gerber/gerber.htm

24 P. Marmet - *Einsteins Relativitätstheorie kontra klassische Mechanik* Vol. V *Berechnung der Drehung des Perihels von Merkur.*
    https://www.newtonphysics.on.ca/einstein/relativitaet05.pdf and
    https://www.newtonphysics.on.ca/mercury/merkur_v2.pdf

sume an irrotational vortex for the planetary system, the observed natural rotations only fit our assumption for Jupiter and Saturn.   The character of a rigid-body vortex increases within the orbit of Jupiter. Between Jupiter and Mars, the shear forces for large planets could have become too great to exist in an irrotational vortex. Mercury and Venus clearly show rotational properties that indicate a rigid-body vortex, while Earth and Mars move in a transition phase between the two types of vortices. See Table 1

Planet	Distance in AE	Orbital speed in km/s	Self-rotation in relation to the sun in km/s
Merkur	0,39	47,40	0,00
Venus	0,72	35,20	0,00
Erde	1,00	29,80	0,46
Mars	1,52	24,10	0,25
Jupiter	5,20	13,60	12,38
Saturn	9,53	9,68	9,53
Uranus	19,33	6,81	2,55
Neptun	30,00	5,43	2,69

Table 1

4.2 Is a Galaxy a Forced Vortex and What Does Force a Galaxy?

According to the doctrine, a galaxy is a large collection of stars, planetary systems, gas nebulae, dust clouds, undefined *dark matter* and other astronomical objects with a total mass of typically 10^9 to 10^{13} solar masses bound by gravity.

The problem is that the measured rotational speeds across the spiral galaxy profile do not correspond to an irrotational vortex according to Kepler and Newton. They lack the mass to generate such rotation profiles. Instead of focusing on physics, however, astrophysicists borrow from the imagination. Even a halo of exotic dark matter is not enough for them.

Fig.15: Galaxy - Fantasy and Reality

The latest academic studies assume that in the center of every galaxy there would be a supermassive *black hole*, which should be significantly involved in the formation of the galaxy, although black holes are supposed to swallow and compress masses.

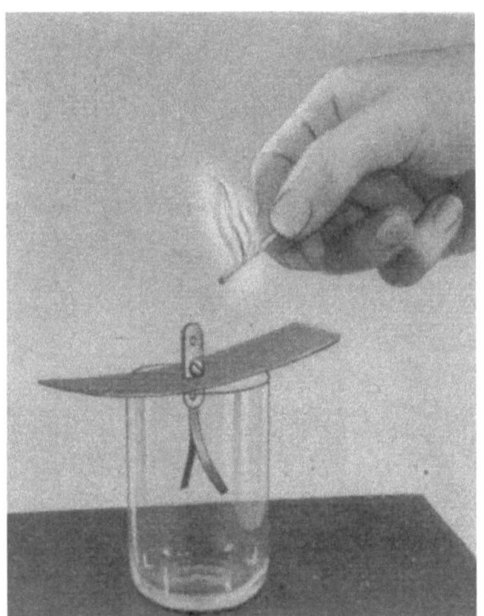

Fig.16: Fire conducts electricity

That all sounds very mysterious and unusual. It is the stuff of dreams, but has nothing to do with physical knowledge. If we compare the representation of a black hole with a real galaxy, we find more light in the center of a galaxy than in its peripheral areas. But what does light mean in the center of a galaxy?

There is a very simple experiment to answer this question that I did when I was 12 years old. I built an electroscope out of two aluminum strips and a holding device, as Fig.16 shows.

I charged the electroscope with a decelith board that I had previously rubbed with a silk cloth, which was visible from the spreading of the aluminum strips. As soon as I approached the electroscope with a flame, I watched the aluminum strips collapse.

As a result, hot gases must dissipate electricity. Why should this experiment only work under terrestrial conditions?

On the contrary, matter in the plasma state, as presented as fire, consists of free charge carriers. The glow is the result of their recombination. There is no reasonable reason to doubt this fact in galaxies, even if prominent astrophysicist Stephen Hawking believed that gravity shaped the structure of the universe. No, my youthful experiments have convinced me that it is the free charges in the gravitational field of the bound charges that have a lasting impact on the structure of the cosmos.

Especially not a gravity, as Albert Einstein understood it, as a space-bending force as it results from singularities of mathematical equations that have no relation to reality, but arise from his pure fantasy and merely satisfy peoples need for the unusual and mystical.

But let us turn back to reality and consider the largest visible dynamic structures of the cosmos. There is now the **S**loan **D**igital **S**ky **S**urvey project *(short SDSS project),* which aims to map the entire visible sky. The database is available to anyone interested. In a project with the aim to classify galaxies according to the Hubble system, I learned that it is not the external shape that is decisive for their classification, but the light spectrum that a galaxy emits. Three classes of galaxies result from the hydrogen content: Young activ galaxies, mature and old galaxies.
In spiral galaxies, for example, the proportion of atomic hydrogen and ionized gases such as oxygen or nitrogen is very high, while in structure-less galaxies the proportion of these gases is most very much reduced. I reported about it in my book *Modern Astrophysics meets Engineering*[25]). From this fact we can deduce that with what was said in Section 3.3, a spiral galaxy, especially the bar-shaped spiral galaxy, can be understood as a rigid-state vortex. However, a rigid-body vortex must be driven by an internal

---

25   M. Hüfner – *Modern Astrophysics meets Engineering;*
     https://www.bod.de/buchshop/modern-astrophysics-meets-engineering-mathias-huefner-9783751920186

force such as a motor. As soon as the inner force is no longer present, it will dissolve and turn into an irrotational vortex.

What do you think will drive this engine? The spectra in the SDSS database, say it must be the hydrogen that generates the electrical power for the magnetic fields to power the galaxy. That will be the theme of the fifth chapter.

Now let's look at how Newton's gravitational equation results in the velocity profile of a bar galaxy without magic and additional exotic cosmic fantasies.

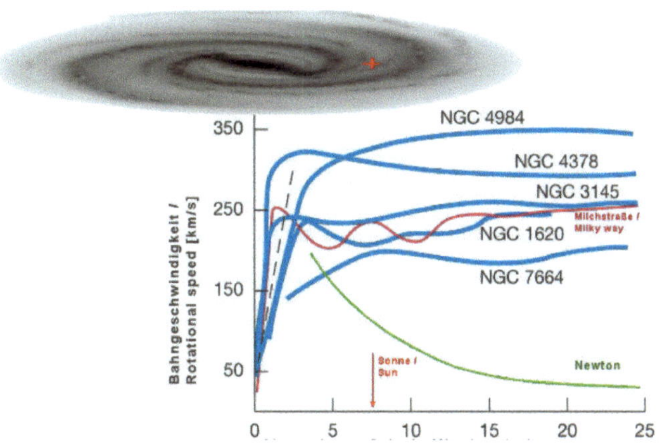

Fig.17: Rotation speeds of galaxies according to Folz &   Eckardt
https://www.atomicprecision.com/Numerical/Paper238b-de.pdf

Figure 17 shows the rotation speeds of spiral galaxies as a function of the distance from the center (blue and red curves). For comparison, the green curve shows the speed distribution that would result after an irrotational vortex according to Newton.

Newton's law not only finds its limit where the two mass radii are comparable with the distance between them, but also where many point charges interact, as shown by the stars in the galaxies, because according to Kepler, the rotation speeds should increase with the distance from the center of rotation decrease indirectly proportional to the square, instead the curves remain almost constant over the entire galaxy disk after a steep increase. While Kepler's observation was valid for two mass points, we are dealing here with a very large number of mass points. Consequently, we cannot expect the speed distribution of a galaxy to behave like two mass points very far apart.

Instead of assuming a halo of mysterious dark matter for this speed distribution, we imagine breaking down the entire mass M in Newton's equation into smaller and smaller components, which are distributed over a cylindrical volume. The more the mass is divided, the more the ratio of surface to volume shifts in favor of the surface. The result is that the free surface charges get the upper hand over the bound charges of gravity. This division is continued until the space between the mass M and the sample mass m is evenly filled with smaller and smaller balls. M is then distributed over a cylindrical disk with the radius r and the thickness d and the distribution density of the mass or the surface charge determines what happens. We get a forced i.e. a rigid-body vortex.

Then the mass of the galaxy is the product of the volume V and the mean distribution density ρ_M over the radius; and the volume results from $V = 2\pi r^2 {\cdot} d$, from which it follows $M = 2\pi r^2 {\cdot} d \cdot \rho_M.$ We can assume that during the process of the crumbling of the mass, free charges do build on the surface of the plasma spheres, which increase with the sum of the surfaces, otherwise the galaxy would not shine. Hence the charge density ρ_Q is proportional to ρ_M. If the test charge m_Q then sits on the surface of the border of the disk V, its distance from the center is also r. We put the extracted expression for M in relation (1) and get

$$|F| \propto 2\pi \cdot d \cdot \rho_Q(r) \cdot m_Q \qquad (5)$$

Because the force is the product of mass and acceleration, the radial acceleration is then proportional to the radial density distribution and the radial force towards the center is canceled out with the centrifugal force. Hence the acceleration

$$b = \frac{\partial v_r}{\partial t} + \frac{\partial v_\theta}{\partial t} + \frac{\partial v_z}{\partial t} \propto d \cdot \rho_Q(r)$$

It follows from this:

$$v_r + v_\theta + v_z \propto d \cdot \rho_Q(r) \qquad (6)$$

In this simple galaxy model, the radial velocity at a location in the galaxy is proportional to the mean charge density of the galaxy and this must be approximately constant, as the velocity plateaus of the blue curves illustrate in Fig.17 using simple math. From this it can be concluded that, conversely, the charge density in the center of a galaxy decreases sharply, which directly contradicts the idea of a gravitational monster in the center of a galaxy. Since mass and charge cannot disappear, a strong current in the z-direction is to be expected as the cause of the lower mass density in the center of the galaxy, and because a magnetic field is present, with a charge separation along the z-axis.

In fact, we now know that mass jets from the centers of rotation occur in practically all spiral galaxies, but this has nothing to do with a black hole in their centra. It is simply due to the conservation of mass and energy, a fundamental law of physics that in 1847 Hermann von Helmholtz already did formulate. And we can derive something else from relation (6): The surface charges of the particles in the galaxy hold the plasma together almost like a solid.

So far we have considered the charge Q of a weighable mass in our model to be at rest. According to Carl Friedrich Gauß, the charge Q is the source of an electric field. When charge moves along the z-axis, we get a magnetic vortex filament along the ionized mass flow, which was first discovered by Kristian Birkeland while studying the northern lights at the end of the 19th century and was named in his honor. These vortex threads take on the function of our earthly power cables in the cosmos.

If the magnetic field changes, an electrical vortex field is emitted, which in turn generates a magnetic vortex field and so on. In other words, when weighable charged masses move, an electromagnetic pulse is emitted, which in turn excites weighable masses in the vicinity and so on is transported, since all masses are force-coupled and bound charge carriers.

Now it is no longer the mass density, but the charge density of free charge carriers that is decisive for further consideration. Many charged particles moving along the z-axis result in a current $\vec{I}$ in the z-direction. This current is characterized by its electrical charge density, not so much by its mass, and it induces a magnetic field that sets all particles in this field in rotation. We are now changing from the mechanical to the electrodynamic approach, because we have recognized electrodynamic forces as the cause of celestial mechanics.

Our guess is

A spiral galaxy is a huge electromagnetic force vortex

It works like a homo-polar motor.

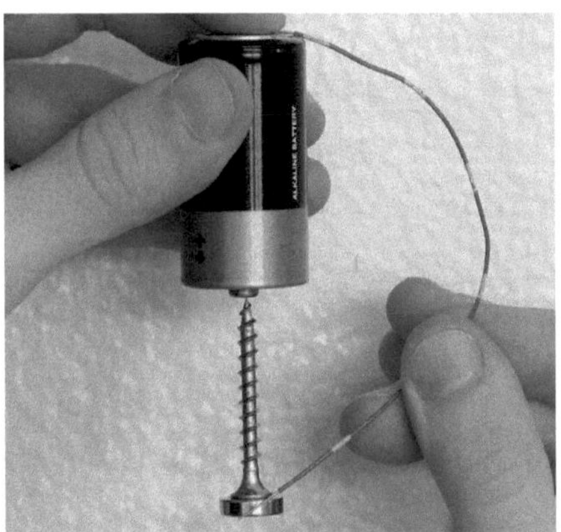

Fig.18: Homo-polar motor-experiment
Photo: Windell H. Oskay

In the next section we want to investigate the structure of the magnetic field of this Birkeland current around the galactic vortex filament, which takes on the function of the copper wire as in our experiment from Fig.18.

4.3 A Force-Free Magnetic Field according to Scott

In Fig.17 we have seen a wavy plateau of the velocity distribution from the Milky Way, our home galaxy (red line). Understandably, the profile is better resolved here than in the distant galaxies. Donald Scott[26]) has developed a more sophisticated model for this, which is to be presented here in its basic features, although it contains somewhat more sophisticated mathematics too.

Let us now consider a current $\vec{I}$ of moving charged particles in a plasma that is not exposed to any external forces. A useful mathematical idealization of such a physical cosmic current is a vector field of current density j which, when viewed in a cylindrical coordinate system, generates an average current vector $\vec{I}$ everywhere, which by definition flows in the direction of the z-axis. The strength of $\vec{I}$ is assumed to be independent of the z coordinate everywhere.

The basic structure of such a cosmic magnetic field is described by the momentum equation of ideal magneto-hydrodynamics.

$$\left(\operatorname{curl}\vec{B}\right)\times\vec{B}=\mu_0\cdot\operatorname{div}\vec{p} \tag{7}$$

Here μ_0 is the permeability of free space. The left side of this expression represents the compressive magnetic Lorentz force and the right side is the expansion force (pressure gradient multiplied by the permeability of the plasma). We differentiate between force-free fields with the partial derivatives $\operatorname{div}\vec{p}=0$ and pressure-balanced fields with $\operatorname{div}\vec{p}\neq0$. Let us consider the force-free case here.

---

26 D. Scott - *Birkeland Currents: A Force-Free Field-Aligned Model*
      http://www.ptep-online.com/2015/PP-41-13.PDF

Then the following applies to the electromagnetic force which every charge experiences within such a plasma:

$$\vec{F}=q\left(\vec{E}+\vec{v}\times\vec{B}\right) \tag{8}$$

$\vec{F}$ in relation (8) is named *Lorentz force.* The first term on the right side $q\cdot\vec{E}$ is the *electrical force* and the second term $q\left(\vec{v}\times\vec{B}\right)$ is called the magnetic force. Here q is the charge and $\vec{v}$ is its flow velocity.

The plasma area receives a cylindrical current flow. Scott does made no initial assumptions about the distribution of the current density over the cross-section. A flow of charge creates its own magnetic field through which the charge flows. The point at which each charged particle q is located is the point of origin of two local vectors in the current: the *current density* $\vec{j}=q\cdot\vec{v}$ and the *magnetic induction* $\vec{B}$. At every point of a magnetic vortex $\operatorname{curl}\vec{B}$ given by Maxwell's equation will be generated a current density vector $\vec{j}$ represented by a charged particle and by the way changed the electrical field.

$$\operatorname{curl}\vec{B}=\mu\left(\vec{j}+\epsilon\,\frac{\partial\vec{E}}{\partial t}\right) \tag{9}$$

The current is called the *displacement current.* It is often considered a null value, as Scott does here, when it can be assumed that there are no time-varying electric fields in the region. The integration of the $\operatorname{curl}\vec{B}$ vector over a cross section of the cylindrical flow (Stokes theorem) gives

$$\int_S \operatorname{curl} \vec{B} \cdot dS = \int_S \mu \vec{j} \cdot dS = \oint_C \vec{B} \cdot dl \qquad (10)$$

where S is any cross-section of the plasma and μ is the permeability or *permittivity* of the plasma medium. The second term in (10) is equivalent to $\vec{I}$, where $\vec{I}$ is the total current carried by the plasma current. We assume with Scott the current cross-section to be circular with the radius r. Then the last term in (10) is $2\pi r \vec{B}$, where the right hand rule[27] applies to $\vec{B}$. Thus the B-field is generated by a cylindrical plasma with its outer boundary, $r = R$, for which is

$$\vec{B}_\theta = \frac{\mu}{2R}\vec{I} \qquad (11)$$

The integral forms of Maxwell's equation given in the both equations (10) and (11) imply that $\vec{B}$ is a vector sum of the effects of all charged particles with $\vec{j}$ vectors on the surface S which are enclosed by the cylinder C below the orbital integral. $\vec{B}$ is not generated directly from a single $\vec{j}$. In (9) it is clear that $\vec{j}$, the current density at a point, only produces a single $\operatorname{curl} \vec{B}$ vector, not a $\vec{B}$ vector of the particle stream. In general, there can be a non-zero $\vec{B}$ vector at points where uncharged particles with $j = 0$.

Before a cosmic particle current system devoid of external forces or fields reaches a stationary configuration, the $\vec{j}$ and $\vec{B}$ vectors interact - all j's generate $\operatorname{curl} \vec{B}$ vectors that add up to form the local $\vec{B}$ vectors. At any point in the plasma where $j \neq 0$ there can be a force between this current density vector and its local magnetic B-field vector. This force is a magnetic Lorentz force given by $q(\vec{v} \times \vec{B})$. This vector cross product of the veloc-

---

27 When the thumb is pointing in the direction $\vec{I}$ along the z-axis, the curved fingers are pointing in the direction of the magnetic field $\vec{B}$

ity vector $\vec{v}$ of a moving charged particle and the local vector $\vec{B}$ means that the scalar magnitude of the resulting Lorentz force $\vec{F}_L$ is applied to each q is given by

$$\vec{F}_L = q \cdot \vec{v} \cdot \vec{B} \cdot \sin\varphi \qquad (12)$$

where φ is the smallest angle between the velocity vector $\vec{v}$, the and $\vec{B}$ vector, with magnitudes v and B. We call φ the *Lorentz angle.* When this angle is zero or 180°, the Lorentz magnetic force $q(\vec{v} \times \vec{B})$ disappears at that point.

The magnetic induction $\vec{B}$ of such an ion current is analogous to a magnetic coil

$$\vec{B} = \frac{\mu \cdot N}{l} \cdot \vec{I} \qquad (13)$$

N is the number of turns and l is the length of the ion current.

This expression shows that $\vec{B}$ is the result of the total current $\vec{I}$. The total energy of the ion current results in analogy to a coil as

$$E = \frac{1}{2} L \cdot I^2 \qquad (14)$$

L is the *inductance* of the ion current. This shows that the only way to reduce all of the stored energy to zero is to completely cut off the flow of electricity.

However, we must assume that the ion current can move and distribute freely in an unrestricted plasma in cosmic space in order to minimize the potential energy stored internally due to the voltages created by magnetic Lorentz forces everywhere in the plasma. In fact, space plasmas are uniquely positioned to obey the **principle of minimum total potential energy**[28]), which states that a system or body must shift and / or deform to a position that minimizes its total potential energy.

The energy described in equation (14) cannot be reduced because it is caused by the fixed current quantity I. Scott argues, however, that the Lorentz energies can be eliminated because they do not depend on the value of $\vec{I}$, but only on the cross products between the local $\vec{B}$ and $\vec{j}$ vectors. As soon as the process of shedding the internal magnetic force-energy reaches a steady equilibrium, this structure is called a *force-free current* and is defined by the relationship between the magnetic field vector $\vec{B}$ and the current density vector $\vec{j}$ at every location where a charge q is present in the effective current:

$$q \cdot (\vec{v} \times \vec{B}) = \vec{j} \times \vec{B} \tag{15}$$

From equation (15) it follows that the Lorentz forces are equal to zero everywhere in a force-free stream (see 12), since every $\vec{j}$ is collinear with its corresponding $\vec{B}$. This arrangement is therefore also referred to as *field aligned current* (FAC). From Maxwell's equations (9) and (15) it follows directly, if there is no time-varying electric field, that (15) is equivalent to

$$(curl\,\vec{B}) \times \vec{B} = 0 \tag{16}$$

what is identical to $\mathrm{div}\,\vec{p} = 0$. This is the fundamental defining property of a force-free, field-oriented flow, which is also known as *weightlessness.*

---

28 H. Callen *Thermodynamics and an Introduction to Thermostatistics,* 2nd ed. John Wiley, New York, NY, 1985.

It also follows from expression (9) that if at any point in an otherwise field-aligned stream $j = 0$, then condition (16) is automatically satisfied even if B is not zero. The value of the magnitude and direction of $\vec{B}$ at a given point is generally insufficient information to determine the magnitude, direction, or even existence of $\vec{j}$ at that point.

This was the problem with which Kristian Birkeland was confronted at the turn to the 20th century by his attempts identifying the currents of the solar wind that were responsible for the magnetic field fluctuations. However, from Maxwell's equation (9) we know the direction and size of the vector at a given point, which there is identical to the value of $\mu \cdot \vec{j}$.

Field-aligned, force-free currents represent the lowest level of stored magnetic energy attainable in a cosmic current [29]).
Let us see now how Scott was looking for an expression for the magnetic field B (r; θ; z) in such a field aligned current structure.

4.4 Scotts Model of a Field Aligned Current

Because $\left(curl\,\vec{B}\right)\times\vec{B}=0$ is satisfied when the current density $\vec{j}$ has the same direction as $\vec{B}$ regardless of its magnitude, it has been proposed by Stig Lundquist [30]) and others

29   A. Peratt - *Physics of the Plasma Universe.* Springer-Verlag, New York, 1992, p. 44. Republished ISBN 978-1-4614-7818-8, 2015, p. 406.
30   S. Lundquist - *On the stability of magneto-hydrostatic fields.* Phys. Rev., 1951,

$$curl\,\vec{B}=\alpha\,\vec{B} \tag{17}$$

which according to (16) is equivalent to

$$\mu\,\vec{j}=\alpha\,\vec{B} \tag{18}$$

where α is a scalar other than zero according to (17). This leads to a simple solution, but from the outset it is important that for any value $\alpha \neq 0$ a non-zero value B at any point requires the presence of a current density $j \neq 0$ at the same point, which is generally an unjustified guess. This is particularly true in view of the known tendency of plasma to form filaments (creating areas in which $j = 0$, but B does not). However, Scott is looking at this special case.

Now you can express the left side of equation (18) in cylinder coordinates:

$$curl\,\vec{B}=\left(\frac{\partial B_z}{r\,\partial\theta}-\frac{\partial B_\theta}{\partial z},\frac{\partial B_r}{\partial z}-\frac{\partial B_z}{\partial r},\frac{\partial\left(rB_\theta\right)}{r\,\partial r}-\frac{\partial B_r}{r\,\partial\theta}\right) \tag{19}$$

and the right hand side of (20) is then:

$$\alpha\,\vec{B}=\left(\alpha\,B_r,\alpha\,B_\theta,\alpha\,B_z\right) \tag{20}$$

In (19) and (20) all field components are functions of the position of $\vec{p}$. Since there is no reason to assume a variation in the current density j in the θ or z direction in cosmic space, equation (22) suggests that this also applies to B. Since no external forces act, except possibly a static axial electric field to maintain the current $\vec{I}$ by $q{\cdot}\vec{E}$, follows that all partial derivatives of $\vec{B}$ with respect to θ and z are zero and therefore from (17) after these simplifications in (19) the following three expressions remain:

Vol. 83 (2), S.307–311. Available online:
http://link.aps.org/doi/10.1103/PhysRev.83.307.

1. In the radial direction is

$$\alpha B_r = 0 \tag{21}$$

There is no radial component of the B-vector. This agrees with Maxwell's $\operatorname{div}\vec{B}=0$.

2. In azimuthal direction we get

$$\frac{\partial B_z}{\partial r}=-\alpha B_\theta \tag{22}$$

3. and in the axial direction we have

$$\frac{\partial\left(rB_\theta\right)}{r\,\partial r}=\alpha B_z \tag{23}$$

This leads to two non-trivially coupled differential equations in the two dependent variables B_z and B_θ, as shown in (22) and (23). The independent variable in both is the radial distance r.

The combination of (22) and (23) gives a second order differential equation in a single dependent variable.

$$\frac{\partial^2 B_z(r)}{\partial r^2}+\frac{1}{r}\cdot\frac{\partial B_z(r)}{\partial r}+\alpha^2 B_z(r)=0 \tag{24}$$

The dependent variable $B_z(r)$ is the axial component of the force-free stationary magnetic field. The component field $B_z(r)$ may extend as far as the current extends. No constraint is introduced for a non-zero value of r. In all real currents in space there is a natural limit $r = R$ for the extent of the current density $j(r)$.

After the differential equation (24) has now been fully specified, we recognize in it the Bessel equation with scalar parameters and, according to the textbook, we get the solution:

$$y=AJ_0(x)+CY_0(x)$$

$J_0(x)$ is the Bessel function of the first type and zero order, and $Y_0(x)$ is the Bessel function of the second type. The former has the value 1 at the limit $x = 0$, and the function $Y_0(x)$ has a singularity on the same boundary. Since reality dictates that the magnetic field remains finite, the value of the arbitrary coefficient C must be set equal to zero. Thus Scott obtains the solution to (24) given by:

$$B_z(r)=B_z(0)J_0(\alpha \cdot r) \tag{25}$$

and

$$B_\theta(r)=B_z(0)J_1(\alpha \cdot r) \tag{26}$$

This Bessel function of the first type and the order zero is used to generate the Bessel functions of the first type and the orders 1, 2, 3, ... by simple differentiation. The recursion formula for the first-order Bessel function is

$$J_1(x)=\frac{d\,J_0(x)}{dx} \tag{27}$$

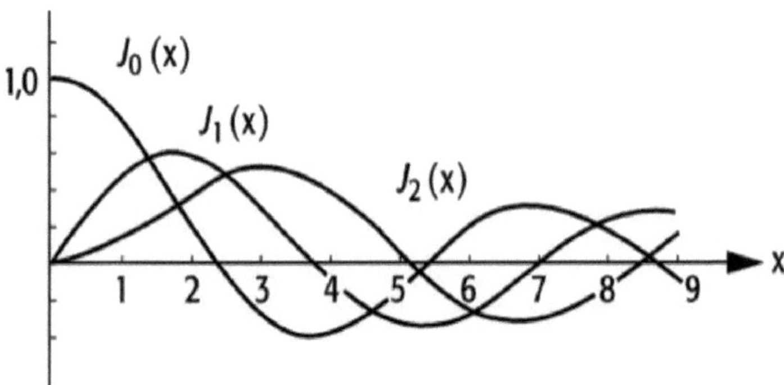

Fig.19: Bessel functions from zero to second order

The Bessel function model Fig.19 of a force-free current explicitly includes only two canonical variables: the magnetic field *B(r)* and the electrical current density *j(r)*. The model requires that these two vector quantities, as already stated above, are everywhere parallel (not interacting).

Since we assume that the flow is of unlimited length and circular cross-section, the model does not take into account any variation of *B* or *j* in the *θ* or *z* direction. From Stokes' theorem (13) then follows for the current density:

$$j_z(r) = \frac{\alpha \cdot B_z(0)}{\mu} J_0(\alpha \cdot r) \tag{28}$$

and

$$j_\theta(r) = \frac{\alpha \cdot B_z(0)}{\mu} J_1(\alpha \cdot r) \tag{29}$$

The relationship between current density $\vec{j}$ and flow velocity $\vec{v}$ is general:

$$\vec{j} = \rho_M \cdot \vec{v} \tag{30}$$

Now only the connection between charge density and mass density is missing and we get the current density that Einstein wanted to get out of the Maxwell equations. We can say that the charge density depends on the degree of ionization of the plasma. The degree of ionization can be deduced from the radiation intensity of a galaxy. The radiation intensity is the dissipative entropy emitted by the open system that the Maxwell equations describe as an electromagnetic radiation field. So we can conclude that the rotation speed of a galaxy is proportional to the ratio of current density and mass density, which explains the waviness of the plateau of the speed distribution along the galaxy's radii in Fig.17.

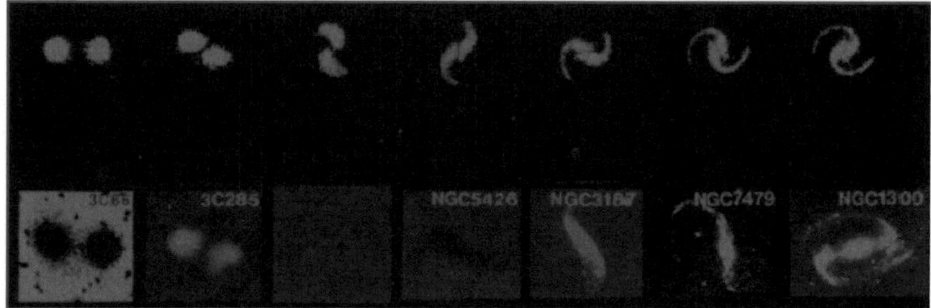

Fig.20: Galaxy simulation compared to observed galaxies
reference: A. Peratt

In fact, the structure of a galaxy is likely to be even more complicated, because these FACs build up even more complicated vortices as soon as $\vec{B}$ and $\vec{j}$ are no longer parallel. as Anthony Peratt[31]) simulated using a supercomputer. We take an important insight with us:

The masses are subject to density fluctuations across the cross-section of plasma currents, as described in the microcosm by a wave function.

---

31 A. Peratt - *Evolution of the Plasma Universe: II. The Formation of Systems of Galaxies;*
https://www.academia.edu/9156671/Evolution_of_the_Plasma_Universe_II_The_Formation_of_Systems_of_Galaxies

4.5 The Spectral Redshift and the Mar of the Expansion of the Space

There is a difference between a physical volume filled with mass and a mathematical relation called space. The *space* is practically the measure with which I measure a volume. If I increase my capacity, I can capture more mass. This is called scaling. It is a question of consideration, not physics, because the total mass is no longer increased. Also, it doesn't make sense to have a measure that expands.

In fact, space and volume are confused in cosmology. It is not the expansion of space that is meant, but that of a closed volume as a function of force when the distances between observer and objects increase. In contrast, an open system you can't blow up. The driving force behind this phenomenon in a closed system is now believed to be the mysterious *dark energy*. If it existed, we would have to be able to prove it on earth as well. A scientific approach, however, means that we prefer natural explanations because the cosmos is as natural a place as our earth.

When we deal with the dynamics in an open cosmos, in addition to the speed of light, the speed of the material luminous bodies is of central physical importance.

The speeds with which we operated in the previous section are determined based on the *redshift* of the spectral lines over the cross-section of the galaxies. A luminous object that moves away from us shows a shift in its spectral lines to the longer-wave part of the spectrum and an object that moves towards us shows a blue shift. This is called the *Doppler effect*. So if the light from one side of the galaxy is red-shifted and blue-shifted from the other side, then you know that the galaxy is rotating. The escape velocity v of the light source with respect to the observer can be determined from the shift in the wavelength.

$$v = \frac{\Delta\lambda}{\lambda}\cdot c = z\cdot c \qquad (31)$$

The Doppler effect results from the addition or subtraction of the speed of light and the amount of the relative speed between the light source and the observer.[32]) Based on the observation that almost all galaxies show a spectral redshift, it appears that they are moving away from the observer, which is why George Lemaître assumed that the volume of the cosmos must expand and the force for this expansion must come from the *Big Bang*. This simple thesis seems plausible at first glance. However, if the spectral redshift is examined more closely, the thesis of volume expansion is no longer tenable. Edwin Hubble, who was the first to determine the spectral redshift as a function of the luminosity of the galaxies, could never agree with Lemaître's thesis, and his student Halton Arp spent his life collecting evidence against Lemaître's thesis.

Even when the sun is on the horizon in the west in the evening, its light is also red-shifted. Nobody assumes, however, that the sun will then move away from the earth and then approach the earth again the next morning. No, when the sun is on the horizon in the evening, it appears larger, i.e. closer than when it is high in the sky. Consequently, it seems as if the light gets tired when it has to cross the thick layer of atmosphere. It loses energy to the atmosphere and with it speed. The spectral redshift can only be

---

32   Einstein's requirement v + c would be c, relates to the use of c as the projection center for the Lorentz transformation. The Doppler effect and the field of optics would not exist if this requirement were fulfilled. We didn't have telescopes

used to a limited extent for the assessment of an escape movement if you have to fear that it will cross cosmic gases.

In the fifties of the last century, a quasi-stellar object (*Quasar*) was found that had a spectral redshift of 0.158, which was far above the redshift observed in galaxies so far. The most extreme recently discovered object of this type was found to have a redshift of $z = 7.642$

This now implies the question of a maximum escape speed from material bodies in general. But now we know that a material body cannot reach the speed of light.

This follows from the law of conservation of momentum and energy. From this, an upper and a lower speed can be estimated according to their masses and a geometric mean can be formed. This gives us the mean limit speed v_{Max} at a little more than 1000 km/s of protons, if we assume that an electron is shot out of its orbit around the proton at the speed of light. In the solar wind they do not reach more than 750 km/s near the earth.

If we estimate the redshift generously, we get a limit for the Doppler effect of:

$$z < \frac{v_{max}}{c} = 0{,}0034 \tag{32}$$

From this it follows that in the case of larger spectral red-shifts, the speed of light in the denominator of the above quotient would have to decrease. This means, however, that there must be gases between the light source and the observer that slow down the wave front of light towards the earth, just as our atmosphere slows down the light when the sun is low in the evening.

Whether the spectral redshift is due to the Doppler effect or the loss of energy in a medium can also be distinguished from the effective line width.

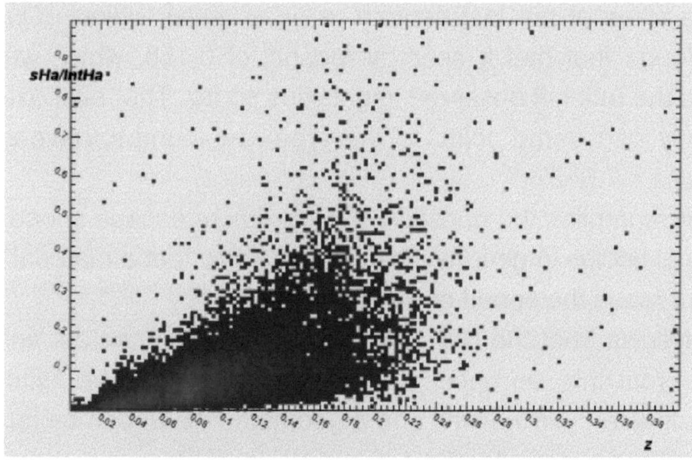

Fig.21: Line width of the hydrogen line as a function of z distance
(each point represents a galaxy)
reference: SDSS-Datenbank

Because with the Doppler effect the light source moves as a whole, the line width must remain unaffected by the environment. It must not have any functional dependence on z.

The line width behaves differently when crossing a gas. When the electromagnetic waves interact with the gas molecules, they will absorb part of the energy, whereby the radiation energy of the wave passing through will be lower. The spectral line decreases in height and shifts its maximum in the direction of longer wavelengths. This gives us a linear dependence of the line width on the spectral redshift. We know this phenomenon as the Compton effect.

100

I examined this connection using the H_α lines of hydrogen. Hydrogen is the most common element in the cosmos and can still be found in all optical spectra despite its spectral redshift. To do this, I filtered out a sample of around 300,000 galaxy spectra from spiral galaxies from the SDSS database. Fig.21 shows my result, a functional relationship between the line width and the spectral redshift, which, given by the Doppler effect, should not exist.

This result shakes the whole academic worldview of the cosmos, because the distance information is based on the spectral redshift. It is therefore not surprising that Halton Arp, with his astronomical observation results, which he published as early as 1985, did not find any recognition among theorists. He wrote 1998 in the preface of his book *Seeing red - Redshift, Cosmology and Academic Science*

> *»Now we have the situation where new facts are judged by whether the fit old theories. If they do not, they are condemned with the judgment: There is no way of explaining the observations, so they cannot be thru. «*[33])

Even 23 years later, the majority of the astrophysicist academic community is still ignorant of the facts and still believes that there is only one explanation for the spectral redshift - the Doppler effect - and thus the ultimate expansion of the cosmos although

---

33   H. Arp - *Seeing red -Redshift, Cosmology and Academic Science;*
https://www.abebooks.de/9780968368909/Seeing-Red-Redshifts-Cosmology-Academic-0968368905/plp

from Fig.21 it can be read that with relation (32) less than 1.5% of the redshift is due to the Doppler effect.

This knowledge makes distance information about galaxies and also age information highly unreliable, especially since Arp identified the quasars as galaxy nuclei that are ejected from galaxies from time to time and their spectral redshift decreases with increasing distance from the parent galaxy which is quite puzzling but will surely find a natural explanation in the future.

> As I write this, the television is showing images of the burning forests in the Mediterranean region and above them a deep red sun behind a shimmer of smoke.

4.6 Black Hole or a Quasar Vortex?

The myth about the black hole began when Karl Schwarzschild solved Einstein's cosmological equation and found a singularity in it, which was interpreted to mean that all mass will be engulfed in and condensed into the black hole. Even the light would disappear in this hole what its name is derived from. Now, mathematicians are sometimes very strange unworldly people. They believe their equations more than the observable nature, although their equations are usually only very crude models of an idealized reality.

This myth of the unusual gravitational monster persists and fuels the fantasies of many pseudo-scientists, although the decades-long argument between Leonard Susskind and Stephen Hawking ended with Hawking declaring in 2014:

> *»I take this as indicating that the anti de Sitter metric is the metric that interpolates between collapse to a black hole and evaporation. There would be no event horizons and no firewalls. The absence of event horizons mean that there are no black holes - in the*

sense of regimes from which light can't escape to infinity. ... «[34]*)*

His sick brain suggested redefining the black hole as a meta-stable limit state of the gravitational field in a hyperbolic five-dimensional space what is actually being tried[35]). Theorists can do that, but they shouldn't dare to want to mix their fantasy worlds with the real three-dimensional world of engineers.

The famous narratress J. K. Rowland had clearly separated the real world from the fantasy world with defined entrance gates, such as at London's Kings-cross station platform nine three quarters.   Using the cosmos as a gateway to the fantasy worlds of crazy mathematicians is becoming increasingly unattractive due to its increasing technical use and represents a huge loss of reputation for the entire academic science.

In the meantime, such pseudo-scientists have apparently conquered chairs at many universities under the protection of the Church, with the consequence that increasingly young people are turning away from mathematics and physics. At least I find worrying this development, as I am observing it in Germany. As a result of such aberrations, science in general no longer provides orientation for many people and I consider that dangerous.

---

34 St. Hawking – *Information Preservation and weather forecasting for Black Holes*; https://www.semanticscholar.org/paper/Information-Preservation-and-Weather-Forecasting-Hawking/d545549ee1b64d234d61b2345330083f7fd849fa

35 V. Netchitailo - *5D World-Universe Model. Space-Time-Energy;* https://hal.archives-ouvertes.fr/hal-02388103

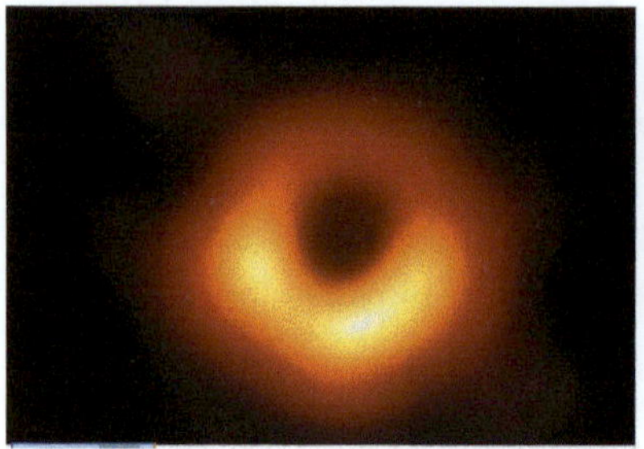

Fig.22: elliptical galaxy M87: left - gas and aerosol vortex in IR; right- with jet stream in visible light;
references: EHT Kollaboration & ESO

An example of the hype that hit the public is the here in Fig.22 seen shot of an alleged black hole. On April 11th in 2019 the German Tagesspiegel headlined:

Like the entrance to hell

> *»With the help of several radio telescopes, it has been possible to depict the center of the galaxy M 87. ... The first real recording of a black hole. «*

We learn:

> *»In April 2017, the astrophysicists aimed eight telescopes - on Hawaii, in Chile, Mexico, Spain, the USA and at the South Pole - at different targets for six days and recorded the radio waves received. The suspected black hole in the center of the Milky Way, also known as Sagittarius A *, and that of the giant galaxy M 87 were observed particularly intensively. ... The recorded signals were evaluated on supercomputers at the MPIfR in Bonn and at the Haystack Observatory in Haystack, Massachusetts (USA) to get that first picture of a black hole.«[36])*

Anyone who paid attention in physics class learned at some point that radio telescopes can receive wavelengths from mil-

---

36 https://www.tagesspiegel.de/wissen/bild-von-einem-schwarzen-loch-wie-der-eingang-zur-hoelle/24203672.html

Fig.23: Jet-Stream out of M87,
reference: NASA team

limeters to meters. These are wavelengths that are also used by amateur radio operators, the propagation of which can be influenced by our ionosphere. This can be used to detect cold gas molecules such as water, carbon dioxide and other hydrocarbons, but also aerosols. Every amateur radio operator has an idea of how such waves are created and that it is so hot in the center of a galaxy that these molecules are broken down into their atoms and ions and smaller structures then also emit shorter waves, for which a radio telescope is blind. In order to make a picture from the radio signals, the wavelength has to be shortened by a factor of 10^6.

But M87 is exactly the galaxy emanating from a jet stream that the NASA team observed in 2017. Now, according to the theory, it has always been propagated that in a black hole all matter including light is swallowed. Here Fig.23 shows exactly the opposite of what astrophysicists have been preaching for over half a century.
A beam of ions shoots out of the center of the galaxy. Oh, now the black hole has just been redefined and all theory remains the same.

At this point we say goodbye to the old academic theory and tear down its walls. We looked at self-similar fractals in Chapter 3 and found that vortices have fractal properties. As a result, we are now looking around whether we know of a vortex device in our scientific world of experience that can produce jets of matter.

The device we are looking for is the *dense plasma focus* (DPF), which was developed in the 1950s by N. V. Filippov at the Kurchatov Institute for nuclear fusion experiments in Moscow. A basic sketch is shown in Fig. 24. The basic principle is the squeezing of a current by strong magnetic fields at the end of two concentric cylinders. The similarity with the production of smoke rings is unmistakable. Only here does the Lorentz force seem to twist and compress the plasma ring more strongly. The energy can be obtained from a capacitor bank and we recognize a plasma focus from a distance through its electromagnetic radiation, which extends into the X-ray range.

In order to identify a DPF principle in the cosmos, it requires electrostatically charged clouds of aerosols, which can be detected with radio waves and the X-rays from the jet stream.

If these aerosol clouds form a ring and form also a double layer, we get a central hollow electrode and an outer ring electrode, as shown in figures 22 and 24.

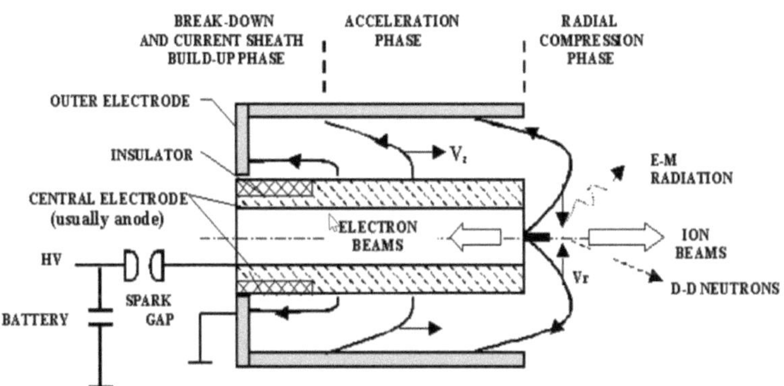

Fig.24: Sketch of a DPF – reference: M Scholz

Now this construction only needs to be charged and a jet stream is created.   This radiation combination of radio waves and gamma rays combined with a strikingly strong redshift of the optical spectrum was discovered in quasars, which, according to the old theory, were suspected to be at the edge of the recognizable world and now move to the center of active galaxies.

What academic science considers a black hole in M87 was discovered by us as a vortex structure. Other galaxies have a spiral structure with and without bars. As far as their development is concerned, we are at the very beginning of understanding and it is not helpful to let our thinking be squeezed into centuries old ready-made theories. On the contrary, we have to be open to alternatives at all times.

In Munich Garchingen Halton Arp dedicated himself to these strange objects, especially in the nineties, supported by the data of the ROSAT satellite, which systematically looked for X-ray sources in the cosmos from 1990 to 1998. He realized that the nuclei of active galaxies have properties like those found in DFPs and that the quasars are quite closely related to such galaxies, although their redshift is much stronger than that of the parent galaxies. He laid down his observations in his book *Seeing Red-Redshift, Cosmology and Academic Science*, arguably the most important work of astronomy in the late 20th century.

As is so often the case in the history of science, he did not receive any recognition from his colleagues for this during his lifetime. On the contrary, he was not infrequently hindered in his work. At the Palomar Observatory near San Diego in particular, he fell out of favor because of his theory about the cosmos. The Max Planck Institute (MPI) in Munich Garchingen offered him a scholarship, which enabled him to continue his work.

The former director of the MPI Rudolf Kippenhahn said about Halton Arp:

> *»We need people like him, otherwise there is a risk that cliques will form in science that do not allow criticism from outside. «*

He remained a dissident and largely unknown in Germany. I regret not having got to know him personally, because when I realized his importance, I could no longer establish contact. because until I realized his importance, I could no longer establish contact. On the occasion of his death, I wrote under the obituary *Halton C. Arp - Death of an Astronomer Who Saw Red* on the website of *RelativKritisch*[37]) as a comment:

> *»All great thoughts appear foolish to the average person because they do not conform to the prevailing doctrine. Neither Galileo nor Darwin were recognized by their contemporaries for their ideas. But has anyone ever seen a botanical treasure discovered on a worn-out street. It is no different with thoughts. What every-*

---

37   http://www.relativ-kritisch.net/blog/kritiker/halton-c-arp-tod-eines-astronomen-der-rot-sah

one thinks will never be enough to make a discovery. Being called a fool or a crank has to be taken into account when you are convinced of an idea. But if you want to make a career, you should keep marching on the given road. «

In contrast to the parlor scholars of astrophysics, his ideas arose from concrete observations on telescopes on cold, lonely nights.

Let's recap what we learned about galaxies from Halton Arp: Active galaxy nuclei have a striking resemblance to quasars. In addition to plasma jets, larger mass units, which have also been identified as quasars, are ejected from these nuclei. The redshift of the quasars has a periodic quantum-like structure in which certain z-values occur in clusters.

If we compare the properties of quasars with those of a DPF, so they are similar to each other, which is why we can assume that in the center of a galaxy there is a dense plasma focus rather than a black hole devouring mass and radiation in a physically inexplicable way.

Correspondences with the structure of a DPF can be expected because the center of a galaxy does not behave like a potential vortex, but like a rigid-state vortex. Due to their huge surface area, aerosol clouds in the galaxies can bind free electrical charges and thus act like huge electricity stores analogous to our thunderclouds. These clouds should also be responsible for the radio waves and the different spectral red-shifts of neighboring cosmic objects.

This assumption came to me when I was flying over Great Britain in fine weather and was able to observe the smog, which had spread in a height over 5000 m above the land with a reddish shimmering veil towards the horizon. Perhaps this aerosol veil is also more effective than CO_2 for global warming in the northern hemisphere of our earth, because it reflects the thermal radiation in the direction of the earth just like the rain clouds in the night sky protect the earth from cooling.

But what about the *gravitational waves* that are supposed to emanate from a black hole that doesn't even exist? The idea of gravitational waves goes back to Einstein's work from January 31st back in 1918.[38]) There he claims that the considered space-time continuum would differ only insignificantly from the Galilean one. Simultaneously, he sets time as imaginary. But that is the major difference to a physical time. In this way he makes time independent of movement and all movement comes to a standstill because time would be vertical to the way. Then, of course, all vibrations and curvatures in space are also imaginary. In doing so, he opens unnoticed a gate to the world of fantasy. A dork looking for those fantasies in reality!

Everyone has already made the experience that the heavier a mass becomes, the more inert it is and the denser it is, the lower its natural frequency. This fact is also used to measure density. To get an idea of how heavy masses vibrate, here are a few examples: Earthquake waves have a frequency in the range between 0.1 Hz and 30 Hz.

---

38   A.Einstein – *Über Gravitationswellen;*
     http://articles.adsabs.harvard.edu/cgi-bin/get_file?pdfs/SPAW./1918/1918SPAW...
     ....154E.pdf

How big masses swing, is by the example of an 850 m long bridge very well documented too, such as the accident of the collapse of the Tacoma Narrow Bridge on November 7, 1940. Shortly before the collapse, it had reached a vibration frequency of 12

Fig.25: The Tacoma Narrow Bridge shortly before the collapse - photo: US Dept of Transportation

vibrations per minute, or 0.2 Hz. The bridge construction was made of iron and we also expect an iron core for our earth.

Should extremely dense cosmic objects vibrate and there should be something like gravitational waves, we would expect extremely long waves according to our experience, the length of which we would probably not be able to measure with earthly means.

However, the LIGO experiment shows us "cirp" signals in Fig.26 with frequencies of 35 to 250 Hz, which do not even rise above the background noise. These vibrations must therefore originate from a much less dense object than a black hole.

What is particularly spicy about this experiment is the fact that an incompetent Nobel Committee awarded another physics prize in 2017 to people who either don't understand physics or are just cheaters. It is very reminiscent of Christian Andersen's fairy tale about the emperor's new clothes, when an imaginary theory is supposed to deliver real results

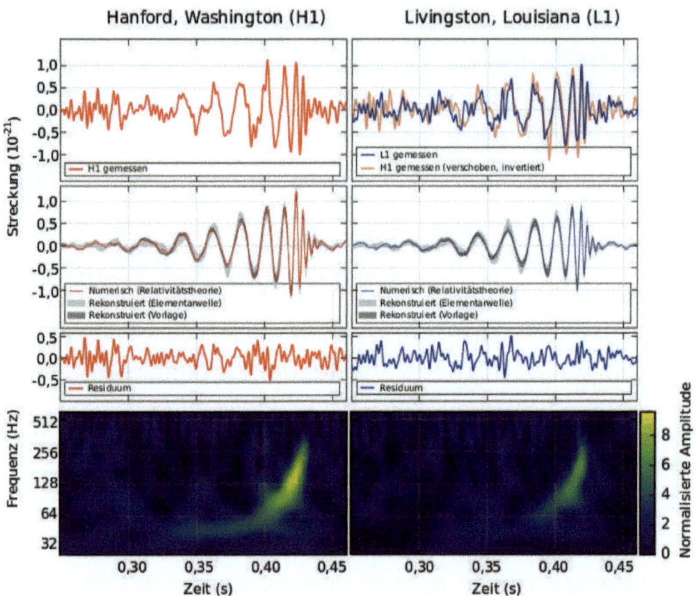

Fig.26: LIGO results on believed gravitational waves

Walter Orlov[39] unmasked this fraud by recognizing these tilting oscillations as an effect that the physicist Heinrich Barkhausen already described in his theory of oscillations from 1932 on page 34. (look at Fig.27)

It is also very bold to assign these tilting vibrations to a cosmic event without being able to establish a real relationship with a cosmic object and without having any physical basis. Orlov recommended that the rather identical spectrometers first be subjected to a thorough investigation with regard to earthly causes for these tilting vibrations.

---

39 W. Orlov - *Zirp von LIGO* http://www.walter-orlov.wg.am/zilch_von_ligo/

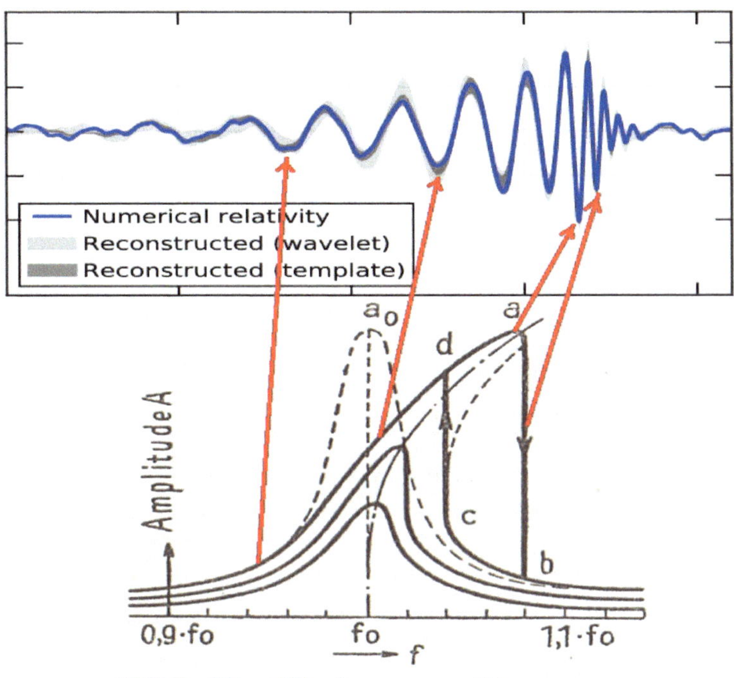

**Bild 25. Verbogene Resonanz-
kurven, Kipperscheinungen.**

Fig.27: Reference W. Orlov

Unlike the macrocosm, the microcosm is usually described in terms of quantum mechanics. Now, however, particles and waves do not want to fit together logically. A wave cannot describe a stable particle either. A wave is a dynamic object made up of many particles. However, if you try to create a particle from the superposition of many different waves, you have to discover that this two-dimensional structure is not stable. It is different with a three-dimensional vortex ring.

In the following chapter we consider the elementary particles as extremely stable vortices of opposite charge due to two mutually perpendicular forces. However, we do not want to speculate about the origin of the load, although a connection with the direction of rotation cannot be ruled out. We accept them as elementary.

5 Structured Atom Models

»The idea of atomic energy is illusionary but it has taken so powerful a hold on the minds, that although I have preached against it for twenty-five years, there are still some who believe it to be realizable.«   - Nicola Tesla 1931

Since the two atom bombs were dropped on Hiroshima and Nagasaki and the subsequent devastating atom bomb tests, it has been believed that the strong and weak nuclear forces in the atom are forces independent of the electromagnetic forces. So far, however, only elementary particles with opposite electrical charges have been found in the atom. Do these forces have something to do with the special structure of the atoms? Let it us to find out.

Anyone who walks through the world with open eyes will notice that nature seems to repeat a tried and tested structure on all levels. But in general it is difficult for physics to describe structures, as point mechanics has not developed a measure for it. In dynamics, it is the closed vortex that is geometrically described as a helix. We also know a mathematical structure that describes a vortex, that is the rotation operator *curl*. It is shown as a vector perpendicular to the direction of rotation.

It is therefore surprising that there has been an irreconcilable separation of macro and microcosm in physics, which has manifested itself on the one hand through the description of the macrocosm using the theory of relativity and on the other hand through the description of the microcosm using the quantum theory with the spin as an easy number. When we considered the macrocosm in the above chapters without the theory of relativity, we primarily considered a vortex in the *r, φ* plane and then we proceeded to expand the vortex in the *z*-direction. We can then understand the wave function of quantum theory as a projection of the helix filament into the r-z plane. A quantum is nothing but a countable amount, but what is meant is the transition from a continuous way of looking at the macrocosm to looking at microcosmic matter as a discrete amount. But where is the transition from one point of view to the other. For many years, efforts without succes have been made to close the artificially opened logical gap between the two approaches with a theory named quantengravity.

In this chapter we want to extend the vortex idea onto the microcosm, especially since Maxwell did not give any information about the size of the electrical and magnetic vortices. But first we have to deal with what hinders this view.

5.1 About the Building Blocks of Matter

The wave function of quantum mechanics can be understood as a projection of the helix-like movement of a vortex in the r, z-plane. However, the mathematical apparatus with its operators and its eigenvalues for describing the states of motion is rather unwieldy, which gives rise to the risk of misinterpretation of the physical facts.

An example of such a misinterpretation is the *spin*. It means something like rotation, but I haven't found a reasonable physical explanation for it. Rotations are indicated by directions, but the spin is not a vector as the operator curl. Academic elementary particle physics divides its particles into two main groups according to spin, in *bosons* and *fermions*.

- The bosons are supposed to carry the forces. The bosons also include the photon, i.e. the light pulse. They are assigned spin 1. It is strange to call the bosons as particles, because they are quanta of impulses that can only be transmitted in a force field. However, we learned in Section 2.2 that an impulse with constant velocity moves through the ether medium without any force. The forces will only occur at the phase transition to a more difficult-to-penetrate phase when the pulse is slowed down. However, no material particles collect there.

- The fermions with the spin ½ divides the academic particle physics into *quarks* and *leptons*. The quarks are theoretical constructs from which *neutrons* and *protons* are supposed to be made, but this has never been proven. There are said to be 6 types of leptons, including three types of *neutrinos*, the natural existence of which is doubtful. Only three observable energy states of the *electron e,* μ and τ remain, of which only *e* is stable. μ has a lifetime of microseconds and τ has a lifetime of less than one picosecond.

Although the spin is physically questionable, it serves the formalism of quantum mechanics and is of paramount importance in particle physics for the classification of its elementary particles.

Its introduction is likely to be subject to some misunderstanding. In the magnetic field it was shown that the optical spectral lines have a fine structure. In 1896 Pieter Zeeman discovered spectral line splits into two or more lines under the action of the magnetic field. So there must still be a feature on the electron that is only visible under the influence of the magnetic field and that is its *magnetic moment.* The obvious conclusion from Maxwell's equations would be that the electron has an elementary magnetic vortex field. Instead, the dubious view has prevailed that it would be a magnetic dipole field.

So in 1921 Otto Stern and Walther Gerlach sent silver atoms through an in-homogeneous magnetic field in an experiment. Since electrons themselves have a magnetic field, they experience a Lorentz force in the in-homogeneous magnetic field and are deflected. Classically, it was expected that the axes of the small magnets could point in any spatial direction and that the atoms would fill an entire area when they hit the screen. In reality, however, only two bands were observed, as if there were only two possible settings. Since silver atoms only have a single valence electron in an s-sub shell and thus no orbital angular momentum and also no magnetic moment caused by it, the only possibility remains that the electrons themselves have their own "vortex momentum" and a magnetic moment associated with it.

Another magneto-electric effect is the Einstein-De Haas[40] effect. It shows that the electrons in a material sample align themselves with the field of a current-carrying coil. As a result, they have to act like rotating an irrotational vortex like we have learned in section 3.3. The mechanical moment of the electron results from this effect, and is identically with the Lorentz-force.

The magnetic moment derived according to Bohr and, on the other hand, as a mechanical moment according to Einstein and

---

40   A. Einstein & W.J. Haas - *Experimental proof of the existence of Ampere's molecular currents in KNAW Proceedings;* 18 I, 1915 Amsterdam, 1915 pp.696-711 https://www.dwc.knaw.nl/DL/publications/PU00012546.pdf

De Haas, provide the remarkable difference of exactly ½. But why this difference, that is illogical.

Since Schpolski[41]) noted that the torsion of the sample in the magnetic field in the Einstein de Haas effect was half a turn and the derivative of Bohr's magneton was calculated with a full turn, the suspicion arises that this difference in the comparison of the theoretical derivation of Bohr's magneton with Einstein and de Haas was not taken into account.

Einstein and de Haas gave a ten percent deviation from the predicted value, which is just acceptable but does not explain the difference between the magnetic and mechanical derivative, which was a factor of ½. There should be no difference at all, because the magnetic moment is the cause of the mechanical moment.

Instead, this factor was given the name spin, although it is obviously just a misunderstanding and has nothing to do with rotation, but rather follows from the different calculation bases for the moments. Schpolski wrote in his Atomic Physics BdII §197:

> *»The idea of electron spin and the peculiarities associated with it were introduced as a hypothesis. As a result, however, it turned out that the existence of the spin and all its properties emerge from the Dirac equation of quantum mechanics, which meets the requirements of the theory of relativity. «*

This is undoubtedly a mistake, since the spin of ½ was entered into the Dirac equation as a premise. There are four coupled differential equations for each coordinate in space-time. You can't get more knowledge out of a model than you put into it. Consequently, the spin is a factor that establishes the relationship be-

---

41   E.W. Schpolski – *Atomphysik BdII;* https://www.amazon.de/Atomphysik-II-Eduard-W-Schpolski/dp/3326000820

tween Bohr's magneton and the mechanical angular momentum for half a revolution. Schpolski himself interpreted the spin as a rotation of the plane of polarization of the wave function of the electron, which we understand as the projection of the helical movement of an electron into the plane.

According to measurements from 2014 the magnetic moment of a single proton is 660 times smaller than that of the electron[42]), but it has also been assigned a spin of ½. The third candidate of fermions, also assigned a spin of ½, is the in-stable neutron. Only the electron and the proton remain of the entire particle zoo as stable particles carrying opposite charges.

If we assume 2.81 fm for the electron radius and 0.84 fm for the proton radius, we get a spherical volume for the electron that is 26 times larger than that of a proton. We cannot know whether the electron and proton are really spherical. Such small structures can no longer be resolved optically. On the other hand, the mass ratio of electron to proton is 1/1836. To get an idea of this, you can compare it to a cloud vortex of one kilogram of water vapor that envelops a luxury car.

So the idea that electron and proton could exist separately in the atomic nucleus, or that they would only be bound to a certain proton in the form of a neutron, is somewhat absurd. Nevertheless, in most cases the electron is optically represented as being small compared to the proton. The proton rather tends to be penetrated by the negatively charged vortex of an electron cloud. The idea of a dipole field for an electron does not fit the Maxwell

---

42   Petra Giegerich - *Magnetic moment of the proton measured with unprecedented precision;* Kommunikation und Presse Johannes Gutenberg-Universität Mainz
https://idw-online.de/de/news590874

equations, which assume a vortex field, and de Climont[43]) has also experimentally confirmed this. A neutron existing outside the atomic nucleus would then be a proton, which would be embedded in an electron swirl.

With a half-life of about 15 minutes, this neutron decays into a proton and an empty electron swirl, which is normally captured by the proton and forced into orbit of its hull and thus converted into a hydrogen atom. The doctrine of quantum mechanics claims that this decay would also produce an additional neutrino. Wolfgang Pauli introduced the neutrino as a hypothesis in 1930 on the grounds of quantum mechanical spin conservation during the decay of the neutron. However, the spin is not a torque with one direction, which is why the spin conservation makes no physical sense and therefore the need for spin conservation is not given.

E.W. Schpolski explains in his Atomic Physics Bd.II that the calculations have shown that a neutrino cannot ionize more than a single pair of electrons per 500 km of airway[44]). However, this means that it is not possible to prove experimentally the existence of a neutrino because no one can guarantee that some other radioactive decay cannot happen on the long journey of that airway.

Nevertheless, huge detectors were built to detect neutrinos. In the Laboratori Nazionali del Gran Sasso at a depth of 1400 m in Abruzzo, around 10 neutrinos are said to have been detected per

43   J. de Climont - *Eine Folge des Rowlandeffekts - Das intrinsische Feld des Elektrons ist kein Dipol* http://editionsassailly.com/drehende_Leitern.pdf

44   E.W. Schpolski - *Atomphysik BdII §288 Das Neutrino;* Deutscher Verlag der Wissenschaften Berlin 1962; https://www.antikvarium.hu/konyv/e-w-schpolski-atomphysik-i-ii-699821

day. The neutrinos are said to have been detected via neutrino-electron scattering in a 300 t unsegmented liquid scintillator. The scintillator called BOREXINO has been collecting data since May 2007. With this detector they want to have directly detected solar neutrinos from the capture of electrons by the transmutation of the radioactive isotope ^{7}Be into ^{7}Li with a half-life of 53 days and an ionization potential of 9.3 eV. To do this, it would first have to be demonstrated that neutrinos are necessary to capture electrons. I have a different explanation for the electron capture in Section 5.3.

Anyone who has already registered atomic decays with an electric tool like a scintillator cannot imagine how can used it to identify neutrinos. But the cost of an experiment of this magnitude cannot be justified if it turns out negative. My suspicion that neutrinos could not be proven beyond doubt was confirmed when I found a paper by the nuclear physicist Carl W. Johnson.
According to Carl W. Johnson[45] the energy that would be transferred to a neutrino would be so small that it would not even be sufficient for the ionization of an ion pair. Johnson writes:

> *Simple math seems to deny that atomic Neutrons could exist inside atomic nuclei. The highly respected NIST database on thousands of atomic isotopes has ten-digit or better accuracy for every isotope Mass. All of the math seems to prove that no Neutron Self-Binding Energy can exist in any atomic nucleus. For example, the simple Tritium version of Hydrogen has a NIST atomic mass of exactly 3.0160492779 AMU (Atomic Mass Units). With a half-life of about 12.33 years, that atom has one of its neutrons beta-decay into a proton and an electron to become a Helium-3 atom which has a NIST atomic mass of exactly 3.0160293201 AMU. The NIST data explains that decay also gives off radiation of exactly 0.0000199578 AMU. Do the math! That simple math where the*

45 Carl W. Johnson - *Nuclear Physics may be Fairly Simple*
    http://mb-soft.com/public4/nuclei7.html

precise AMU of the beginning Tritium atom is still exactly identical to the sum of the resulting Helium-3 atom and the radiation.

Finally, he concludes:

Atomic Nuclei May Not Contain Neutrons or Neutrinos.

Here, too, it has been shown that a spiritual product has materialized over time in a miraculous way according to the principle of the immaculate conception of the Holy Virgin Mary, because the neutrinos are needed to explain how, according to the classical solar model, the sun transports its energy from the interior to the surface without, however, having an explanation for the fact that the solar surface is much colder than the solar corona above it.

5.2 The Electrostatic Structured Atom Model

We learn at school that an atomic nucleus consists of positive protons and electrically neutral neutrons, and around this nucleus there is a negative electron shell that keeps the atom in charge equilibrium. This assumption results from the experience with nuclear fission, which always splits new atomic nuclei as a result of neutrons produced.

The number of protons in the nucleus determines the number of electrons in the shell and thus also the atomic number in the periodic table of chemical elements. The strong nuclear forces are supposed to hold the positively charged protons together, which should fly apart due to the Coulomb force. The weak nuclear forces would be responsible for the radioactive conversion of the elementary particles. A number of core models have been developed over the years. But none of them has yet been able to explain satisfactorily all the properties of atomic nuclei. In addition, none of these models took sufficient account of the internal structure.

This has changed since Johnson's discovery. Instead of neutrons we have protons and also nuclear electrons, which are practically responsible as a cloud of negative charge for the electrostatic cohesion of the protons. This eliminates the need for the core forces. At first glance, that is a very plausible picture. This picture is propagated by Edo Kaal [46]), whom I met personally at the EU2017 in Phoenix, Arizona, which I visited on invitation of the Thunderbolts community. I am supporting this community in spreading the ideas about the Electric Universe (EU).

I was fascinated by the Atom Builder, a computer program that he shared with his colleague James Sorensen had created, with which three-dimensional atomic nucleus models are assembled from spherical protons and their properties can be modeled. In order to understand the matter even more impressively, he used spherical neodymium magnets, some of which he gave me as a gift.
As nice as the model seemed to me, I felt a contradiction between the properties of the magnets and the electrostatic model and why are two different forces observed in the atomic nucleus and how does the model fit into Maxwell's vortex equations? With this silent doubt, I left the conference and played with the magnets until I found another approach.

---

46 https://structuredatom.org/team

5.3 An Atomic Vortex Model

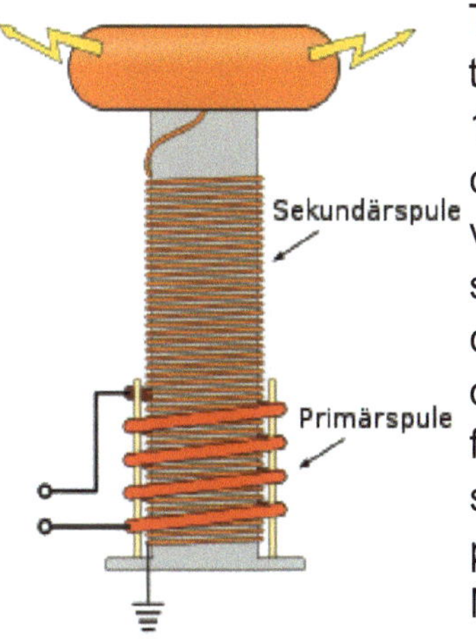

Fig.28: Tesla coil

One day I remembered Nicola Tesla's coil and that this resonant transformer circuit designed in 1891, emits light waves when excited. The principle is simple. Two vortices are arranged around a solid core in the form of a primary coil with a few turns and a secondary coil with many turns. How far can be made smaller such a structure without changing their properties?

Maxwell did not provide any information on this.

A model with a large electric circuit as atomic shell and a small current vortex in the atomic nucleus provides a model with a smaller vortex down to atomic size, whereby the small vortex rotates faster and thus generates more turns. While the primary turn is the electron shell and the secondary turn is formed by the nucleus electrons swirling through a massive nucleus of protons, this nucleus contains about 99.95% of the mass. The primary vortex is three orders of magnitude larger than the secondary vortex with the same angular velocity, which, however, results in a multiple number of turns. As early as 1869, William Thomson proposed to assume that atoms are vortex rings in the ether. Isn't there a fractal resemblance to macroscopic structures?

The English mathematician William Kingdon Clifford[47]) is quoted by Antje Pfannkuchen:

> »The vortex motion alone turns an imponderable ether into ponderable particles of matter; and also the degree of hardness or elasticity of solid bodies would be determined solely by the very rapid motion of something which is infinitely soft and yielding. «

How simple, vortex movements would be the beginning of all things. Perhaps too simple for an academic science?

If you consider the elementary particles themselves as vortices, then the electron and proton form an entangled vortex pair. The interpretation of the quantum Hall effect discovered by Klitzing in 1980 using a torus model by Klaus Gebler confirms this assumption[48]).

The primary vortex is the electron shell and the secondary vortex is formed by the nuclear electrons swirling around a massive nucleus made of protons, which contains about 99,95% of mass. The primary vortex is three orders of magnitude larger than the secondary vortex at the same angular velocity.

Fig.29: Elementary droplet magnet made up of two protons and one electron

Then the nuclear forces would not be electrostatic, but magnetic, because the protons take on the role of the magnetic nuclei. But how do you have to imagine the structure within an atomic nucleus? However, I have to give up the idea that electrons and protons are spheres. To be compliant with Maxwell, let's think of the proton as a doughnut-shaped magnetic vortex. Then the electron is a doughnut-shaped electrical vortex that penetrates the mag-

---

47   A. Pfannkuchen - *Vom Vortex zum Vortizismus; II*
     http://www.newvortex.de/vortex2.html [73] Clifford (1875) S. 784.

48   K. Gebler - *Ein Torus-Modell für Quantenphänomene, dargestellt am Quanten-Hall-Effekt;* unpublished manuscript

netic vortex like a link in a chain. In order to build an elementary magnet, you need two magnetic vortices that are penetrated by an electrical vortex as Fig.29 shows. This structure has two different force value ranges on its surface, on the poles and between.

So I imagined the inner workings of Edo's magnetic balls. An electron vortex be quite large string if you think of it as an unwinded current circuit. The Lorentz force can also tangle a free electron up to the size of the classical electron diameter. It is difficult to imagine that statically repelling electron balls would pull together in an electric field to form a current filament, as the pinching effect shows.

Just as Tesla circuits can contain multiple resonance coils (inductors) I started with this idea to build atomic models out of my droplet magnets. That was easy as long as the atoms were of even atomic number. But what about the odd ordinal numbers? This is where nature comes to my aid. Let's look at the nuclei of the hydrogen isotopes deuterium and tritium. While deuterium is a stable isotope, tritium has a half-life of around 12,3 years. The tritium nucleus with atomic number 1 and two "neutrons", effectively three protons and two core electrons, is converted into a stable ^{3}helium nucleus by releasing one electron. However, ^{4}helium has the atomic number 2 and the Alpha particle is the same nucleus without hull electrons.
The Alpha particle is two times positively charged and the Beta radiation is a negative charged. So we understand the radioactivity as an electrical effect. Why should exist there extra forces in the atom nucleus?

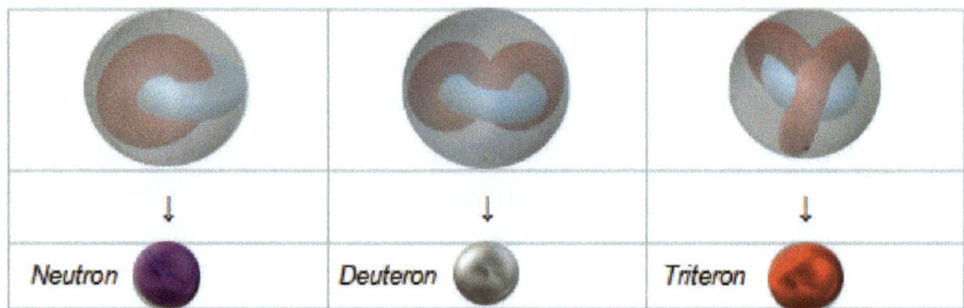

Fig.30: Core modules, The color symbols of these modules are shown in the bottom line.

Fig.30 shows the identified core building blocks in accordance with Maxwell's vortex equations.

For the sake of simplicity we will represent the proton *p* as a red colored doughnut and the *core electron e* as a blue colored doughnut and since the electron appears to be larger than the proton, we will put a spherical shell around one, two or three protons with the *mass numbers M* one, two and three. The mass of the nuclear electrons is irrelevant for our model. The shell electrons are taken into account via the atomic number Z.

The quotient $Q_p = M / (M-Z)$ is important for assessing the stability of an atomic nucleus.

The denominator *M-Z* in the old model is the number of neutrons. Here it is the number of nuclear electrons. It turns out that this quotient should be 2 for a neutral atom in charge equilibrium. If it is greater than 2, electrons are captured to stabilize the nucleus. The electron is taken out of the shell and the atomic number is reduced. The denominator is getting bigger by this. In the event that the quotient is less than 1.85 there is a tendency for an electron to be given off and the ordinal number increases by 1, which reduces the denominator and stabilize the nucleus. But that's not an absolute limit. Small deviations are observed. This shows that the charge on the proton is usually saturated by two electron charges, one from the core and one from the shell. This in turn shows the similarity to a Tesla coil too if you think of the electron vortex as an inductance. On the other hand regular hydrogen has no core electron, so it has the bridge property between two elements.

We can build a neutron model from a red doughnut and a blue doughnut. As you can see, according to the above quotient, a neutron is not at all neutral, but a very aggressive particle, which has the tendency to destroy atomic nuclei and it really does.
But we can also combine two red donuts with a blue doughnut and thus we get a deuterium nucleus model, hereinafter referred to as *deuteron* D for short. This is a stable elementary magnet. But you can also add an additional proton without endangering the stability of this elementary magnet.
Only its magnetic effect on the outside decreases. So you simulate a stable ^{3}He core. We want to simply call the latter component *triteron* T in the future because of the three protons. Although the mass of the nuclear electrons is irrelevant, from the triteron there are a variant with one and another with two electrons, (T_1 a ^{3}He core and T_2 a Tritium core) which are identified by a red with different blue components in the following figures.

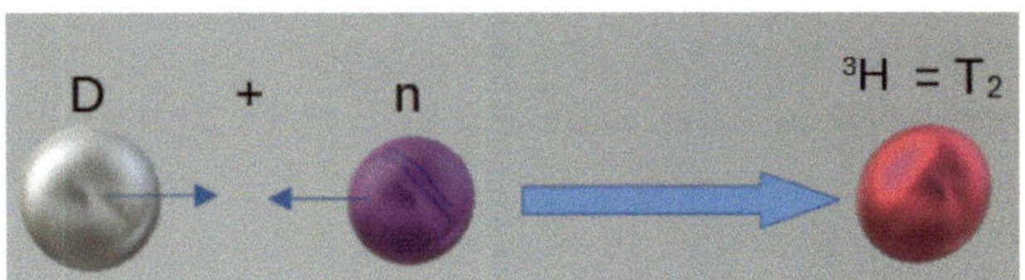

Fig.31: Nuclear fusion

The bottom line of Fig.30 shows the color symbols for the magnetic balls with which the atoms are to be modeled in the following. We can now show the *nuclear fusion* of the first two core modules to the third module in Fig.31.

The latter must have two core electrons. The transmutations formula is

$$(2p +e) + (p +e) \rightarrow (3p +2e) \quad \text{or} \quad D + n \rightarrow T_2$$

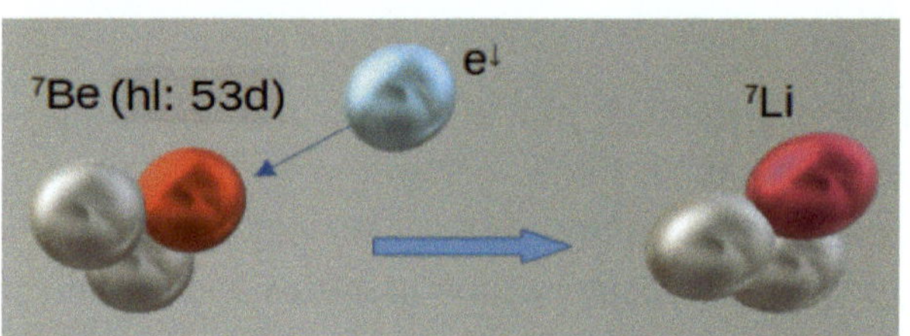

Fig.32: Transmutation with electron capture

When assembling atom nucleus models of higher mass numbers, you do not only get stable atomic nuclei. Atomic nuclei which differ from another nucleus of the same chemical element only in their number of nucleus electrons are called *isotopes*. An other nuclear reaction is the transmutation of the in-stable isotope ^{7}Be with Q_p= 2.33 in the stable ^{7}Li isotope with Q_p= 1.75 in Fig.32. The superscript number in front of the chemical symbol denotes the mass number in order to be able to distinguish the isotopes.

The blue droplet on the left side of Fig.32 symbolizes the electron to be captured from the in-stable beryllium nucleus from its electron shell. Here, too, it can be assumed that the internal electron vortex in the triteron is formed by two electrons. The transmutation formula with electron capture e$^\downarrow$ is characterized by an odd mass number

$$(3p +e) + e^\downarrow \rightarrow (3p + 2e) \quad \text{or} \quad T_1 + e^\downarrow \rightarrow T_2$$

This rule is observed for the first time from ^{7}Beryllium upwards. With an even mass number, another rule from ^{10}Carbon upwards applies

$$2\times(3p +e) + e^\downarrow \rightarrow 3\times(2p +e)$$

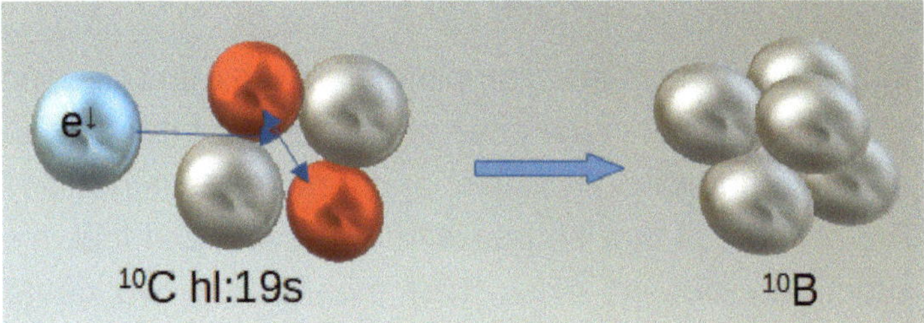

Fig.33: Second transmutation rule for electron capture

Fig.33 shows first time the second transmutation rule for electron capture.

Another rule is the merging of two free deuteron to form an alpha particle. The first time this rule is observed is at ^5Li. The capture of electrons goes hand in hand with the reconstruction of the electron shell and the atomic nucleus, which triggers intense X-rays and gamma-radiation.

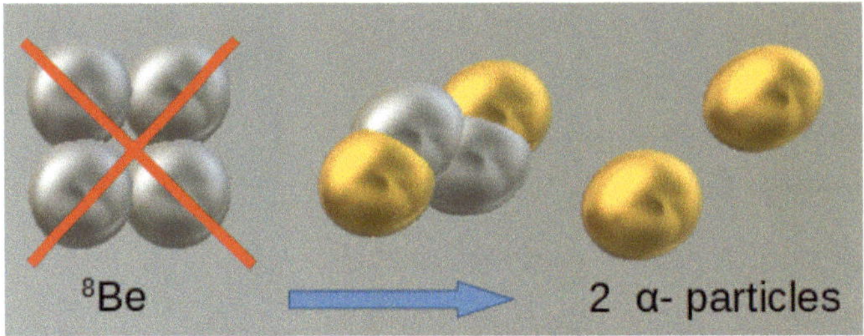

Fig.34: Nonexistence of 8 Beryllium

Another type of *radioactivity* in the construction of atomic nuclei is associated with the escape of particles from the nucleus. We differentiate between Alpha-radiation and Beta-radiation.

The transmutation formula for Alpha-radiation is

$$2\times(2p+e) \rightarrow \alpha$$

Here in Fig.34 we have an example of this rule with two opposing pairs of deuterons, as in the case of the decay of the ^{8}Be isotope into two alpha particles. A doublet or quartet of deuteron isn't stable. The halftime of that isotope is 10^{-16}s what mean that this Isotope practically doesn't exist. This means that two deuterons combine to form an alpha particle when they face each other. As a result, however, they lose their magnetic properties on the outside and thus their cohesion. This is why ^{4}Helium also seems to be so inert. We only observe this rule again at mass numbers greater than 82.

In addition to the already known structure of ^{7}Li from Fig.32, there is also a stable ^{6}Li isotope with Q_p= 2.00 in Fig.33. Both isotopes need 3 electrons in the atomic shell to neutralize the nucleus, but ^{7}Li has one proton and one core electron more than ^{6}Li. This means that at ^{7}Li the charge balance with Q_p = 1.75 is shifted slightly into the negative. This is an effect that is not noticed in the old atomic model.

The second type of radioactivity is beta radiation. Only the ^{9}Be isotope with Q_p= 1.80 is stable, while the ^{10}Be isotope with Q_p= 1.67 releases electrons with a half-life of 2.7 million years and transmutes to boron.

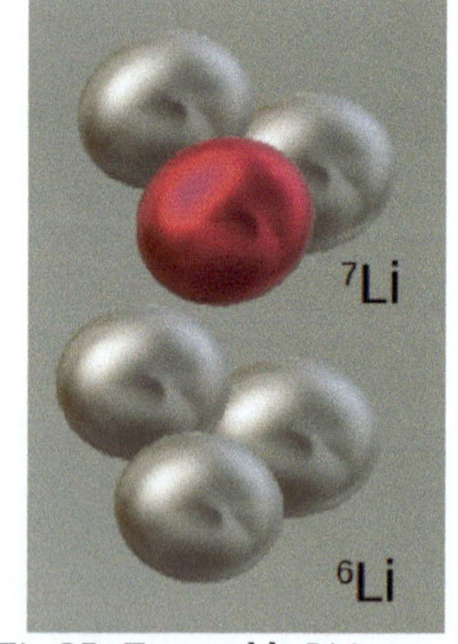

Fig.35: Two stable Li isotopes

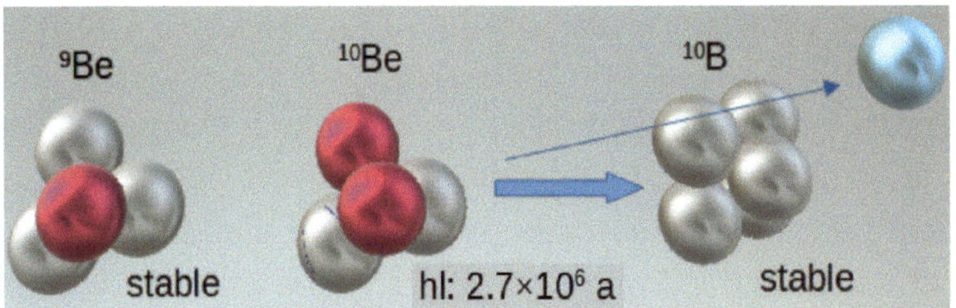

Fig.36: Transmutation of Beryllium

Fig.36 shows a stable beryllium isotope ^{9}Be with Q_p = 1.80 and an in-stable ^{10}Be with Q_p = 1.67, which converts into the stable ^{10}B with Q_p = 2.00 with the release of an electron (Beta radiation). In this model there is no need to introduce a neutrino. The atomic nucleus can absorb the impulse that arises when the electron leaves the nucleus.

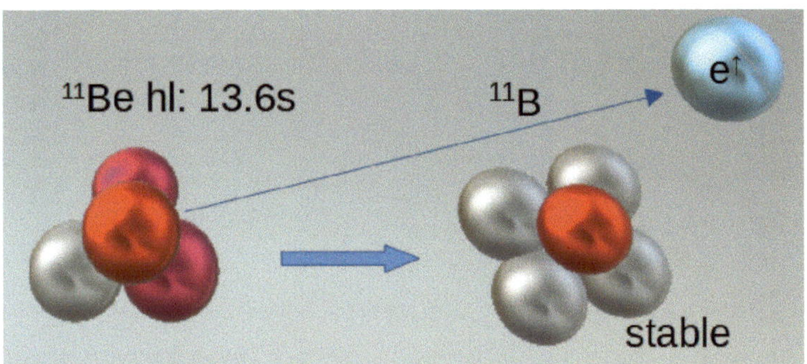

Fig.37: Transmutation of ^{11}Beryllium to ^{11}Boron

The Formula for the transmutation with Beta decay is:

$$2\times(3p + 2e) \rightarrow 3\times(2p + e) + e\!\uparrow$$

Boron has another stable isotope, ^{11}B with the formula $4 \times D + T_1$ with four deuterons and one T_1 building block. It comes from the unstable ^{11}Be as Fig.37 shows, with the formula D + T$_1$ + 2 × T$_2$ and with Q$_p$ = 1.57.

^{11}B with the additional core electron reaches Q$_p$ = 1.83 and is then stable. It has an odd mass number but emerged from the same rule for Beta transmutation as atoms with even mass numbers. T$_1$ is not affected by the conversion.

he next heavier atom is carbon. It has two stable isotopes ^{12}C with Q$_p$= 2.00 and ^{13}C with Q$_p$= 1.86 and one isotope ^{14}C with Q$_p$ = 1.75 and a half-life of 5730 years. ^{14}C is a candidate for Beta-decay. Due to the large distance between the two T$_2$ components, the half-life of 5730 years is enormous compared to ^{11}Be in Fig.37.

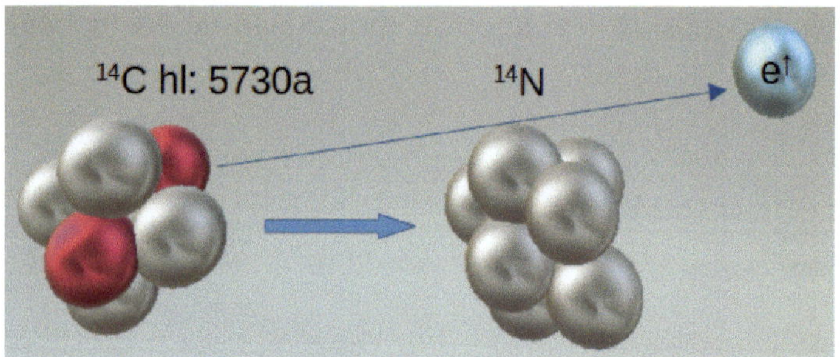

Fig.38: Transmutation of ^{14}Carbon to ^{14}Nitrogen

4×(2p +e) + 2×(3p + 2e) → 4×(2p +e) + 3×(2p +e) + e↑

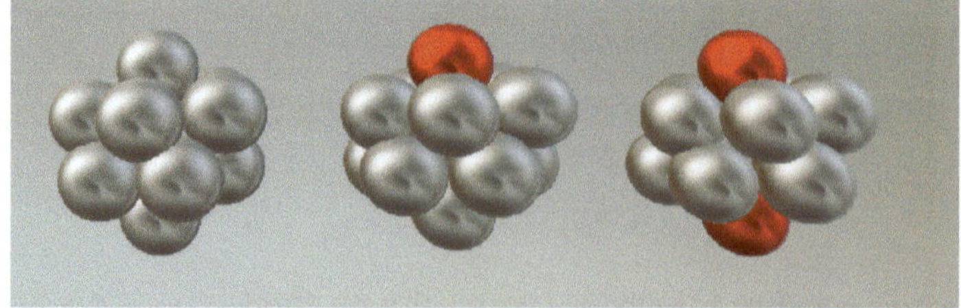

Fig.39: The 3 stable Magnesium isotopes

In Fig.39 you see on the right site two T_1 particles and you would expect that by means of electron capture they would transmute into three deuterons. But because of the distance between them and Q_p= 1.86, that won't happen.

However, with all of our models we never had to resort to the dangerous neutrons.

These examples may suffice to illustrate the transmutation rules. As the number of masses increases, the core structures of the atom models and the number of possible isotopes of the individual elements become more and more complex, so that when studying the properties of these core models with our static display options, we quickly reach limits. One more remark on the limits of the rules. They apply to neighboring particles.

After I had already sketched the concept of a droplet atom consisting of electron and proton vortices in my book *Modern Astrophysics Meets Engineering Sciences*, I found, while studying the history of physics, the vortex theory of Hermann von Helmholtz, based on the first considerations to a mechanical vortex theory of the cosmos by René Descartes. Inspired by Helmholtz, William Thomson, William Hicks and Joseph J. Thomson were already working on the model of a vortex atom in the late 19th century.[49])
While W. Thomson's introduction of vortex atoms had still its starting point in his fascination with hydrodynamics and his critical stance on atomism, later researchers mostly related the vor-

---

49   H. Kragh - *The Vortex Atom: A Victorian Theory of Everything;*
      https://onlinelibrary.wiley.com/doi/abs/10.1034/j.1600-0498.2002.440102.x

tex atom after the discovery of the electron to the development of ideas of electromagnetism, as Edmund Whittaker 1958[50]), Kenneth F. Schaffner 1972[51]), and Howard Stein 1981[52]).

 According to our digital understanding in this days, it is astonishing that there was fierce competition between the representatives of a continuous flow because of the stigmergy of countless particles and the representatives of atomism. There is only one explanation for this: at that time, nobody suspected how many orders of magnitude lay between the two model ideas.

It really pays off from time to time to rethink clever ideas, as suggested by my compatriot Wolfgang von Goethe over 200 years ago and as I put it before the first chapter of this book as my motto.

Nuclear physicists saw the atom from the point of view of destruction. At first atoms were split with terrible consequences in their application and then elementary particles in huge accelerators.

But what are the fragments of elementary particle vortices supposed to represent that form anew in microseconds and what can you do with them? They organized a structural analysis in the form of crash tests and were blind to the extraordinary stability of these elementary vortices, which Helmholtz had already predicted. Have these crash tests in the largest particle accelerator in the world (Large Hadron Collider - LHC) yielded fruitful insights for our society?

I don't see any useful approaches. A vortex produces only an other vortex if Helmholtz is right. Allegedly, after millions of attempts, they found an item named *Higgs boson*, which suppos-

---

50   E. Wittaker - *A History of the Theories of Aether and Electricity*, vol. 1. London: Nelson and Sons.
51   K.F. Schaffner - *Nineteenth-Century Aether Theories;* Oxford: Pergamon Press.
52   H.Stein - " *'Subtler Forms of Matter'* in the Period Following Maxwell'', pp. 309–339 in Cantor and Hodge, *Conceptions of Ether.*

edly transferred the mass to elementary particles. What a nonsense. Elementary means - singular. Mass means - plural. Where did these physicists learn logic?

We are pursuing a different approach here, namely the construction of an atomic order from two differently charged elementary particles. In practice, however, this can only succeed if we are able to remove the entropy from a fusion chamber. So we need to thoroughly rethink nuclear physics for clean energy generation.

5.4 Nuclear Fusion on the Sun

Photosynthesis has stored this fusion energy of the sun in plants on earth and created huge deposits of fossil carbon over many geological ages, which we simply burned wastefully in the industrial age, on the one hand to bring the earth out of thermal equilibrium and on the other hand to process its resources into garbage. This metabolic process has accelerated more and more in the last few decades. Thereby less entropy is removed from the Earth system than is generated internally, which leads to climate change. Politics calls this growth. In the animal kingdom we see that a high metabolic rate shortens the lifespan of individuals. Should societies be able to defy general thermodynamic laws?

We are supposed to believe that only the CO_2 produced by this metabolic process is responsible for the warming. But what about the soot particles? What about the waste heat from the Carnot process, which converts only a third of the energy into mechanical and two thirds into heat? What about land sealing and the deforestation of natural forests our natural cooling system? The future holds many challenges in store for society. The problem from a thermodynamic view point is the acceleration of the metabolism of capitalist society. The life cycles of the products have to be extended and the production routes shortened.

One of the most important tasks of physics in this context is the efficient energy supply of the future. But direct dissipative solar energy is not always and everywhere available in sufficient quantities. Where we have previously used crude oil and natural gas, we want to use hydrogen as an energy carrier in the future. To produce hydrogen we need a lot of electricity to break the water down into its constituent parts. Burning the hydrogen back into water is not an economical process. The energy balance is negative. Even if we obtain all of the energy from solar radiation, considerable areas are lost for natural biological cycles which ensure the reduction of entropy.

After all, we have known since the middle of the 20th century that the sun uses nuclear fusion to generate radiation energy from hydrogen. If we could build an artificial sun, we would be able to use the hydrogen produced more efficiently. But what do we think we know about how this solar energy source works and what do we really know about it?

The academic knowledge is based on a nuclear fusion process that is supposed to exist inside the sun and is supposed to convert hydrogen into helium under high pressure. Although nobody can see into the sun, it is assumed that the sun consists of bowl-shaped zones, some of which can be sharply delimited. A rough division is into the core zone as a fusion furnace, into the inner atmosphere up to the visible surface and above that the outer atmosphere. This fusion furnace would generate the radiation energy in a very slow conversion process by means of the proton-proton chain at temperatures of 15 million degrees and the hypothetical neutrinos would transport the energy to the outside. In the first step, two protons should fuse to form a deuterium nucleus. However, this reaction is very unlikely, on average a proton would take 10^{10} years to react with another proton. This is how they explain the long lifespan of the sun.

Other experiences have been made with hydrogen bombs. It is certainly not a good idea to store the fuel directly in the stove. Hydrogen occurs naturally as a molecule of H_2. Unless it's in an electric field that separates the electrons from the protons. Then two free protons will undoubtedly meet an electron in a fraction of a second, which will combine them to form a deuterium nucleus, if the ambient conditions are right. Since an electromagnetic pulse is observed during a nuclear explosion, you can assume the existence of such an electric field for the moment of the explosion. A thermonuclear reaction is also always an electronuclear reaction too, as we learned in the previous section because it creates a glowing plasma.

Anyone who has paid attention to electrical engineering at school is sure to know a tape generator. Some electrons are brushed off the tape by friction and then you can see a nice flash between its electrodes. Imagine now the discharge, one could remove all electrons from the generator and that in the order of magnitude of the sun. The result would be a supernova. So that's not how the sun can work.

The problem with the standard solar model is that it is a purely mechanical model, which is based on the ideas of Newtonian gravity without electrodynamics and thermodynamics. Elementary particles, however, are electrically charged particles, which is why I always have to take into account the laws of electrodynamics and they move in the plasma state, which is why the thermodynamics of an open system must also be taken into account.

The first electric solar model was developed by Ralf E. Juergens[53] in 1979 after he realized that the calculated energy density of cosmic rays in our galaxy is comparable to the total energy density of electromagnetic radiation including starlight. He recognized the grain structure on the sun's surface as anode tufts and an indication of the stigmergy in an open system. This model was significantly further developed by Don Scott. In 2006 his book *The Electric Sky*[54] appeared.

Scott recognized that the sun forms an anode in a huge circuit connected by *Birkeland currents* and that its energy and fuel are supplied from outside. Hence we have to learn about the galaxy in which a star such as the suns is embedded. In addition to dust, every galaxy also contains huge gas clouds of hydrogen, oxygen and nitrogen around its stars, as can be seen in every galaxy spectrum.

---

53   R. E. Juergens - *The Photosphere: Is It the Top or the Bottom of the Phenomenon We Call the Sun?;*
     https://www.kronos-press.com/juergens/k0404-photosphere.htm
54   D.Scott - *The Electric Sky*
     https://www.amazon.com/Electric-Sky-Donald-Scott/dp/0977285111

There you will find the red H_α line everywhere, as shown by the galaxy spectrum in Fig.40.

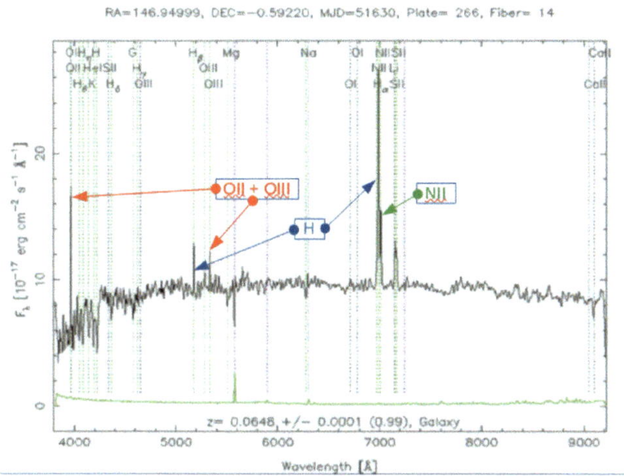

Fig.40: Typical Spectrum of a spiral galaxy

Since the sun's corona is up to one million degrees hot and experiences a steep temperature drop in the photosphere, so that the sun's surface appears almost cold with a maximum of 6000 degrees, it can be assumed that the nuclear fusion does not take place inside the sun, but in its corona. That would also explain the Fraunhofer lines that can be seen in the solar spectrum. These dark lines appear against the background of the photosphere in the chromosphere, where the continuous light energy is absorbed by the newly formed atoms. Fig 40 and 41 show sun's atmosphere and temperature profile.

The hydrogen surrounding the sun is likely to exist as a hydrogen molecule due to its chemical bonding properties.

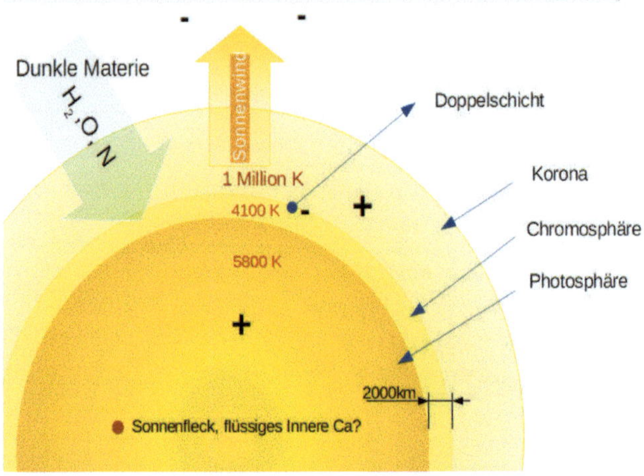

Fig.41: Electrical Sun model

Because the sun's body is positively charged, the hydrogen molecule is attracted to the sun and it loses its shell electrons in the corona. Since the electrons move towards the sun much faster than the protons, there will be an electron deficiency. On the underside of the corona, the protons are strongly decelerated due to the drop in temperature or the increase in field strength. If a current is suddenly slowed down, however, a mass collision occurs and if a few electrons collide with protons, deuterons are created, perhaps also ³helium or tritium nuclei. These then come together according to the rules derived in Section 5.4 and condense to larger nuclei after leaving the corona in the direction of the sun, while the protons, which could not catch electrons, are accelerated in the direction of space as a positive loaded *solar wind*.

The fusion of atom nuclei needs more electrons then the hydrogen can spend. There must be a lack of electrons in this area so that the elementary magnets can organize themselves into larger units.

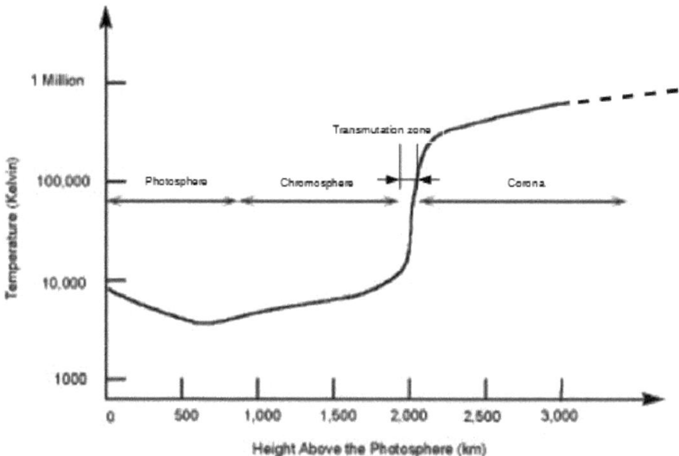

Fig.42: Temperature profile in Suns atmosphere by
http://www.astro.sunysb.edu/fwalter/AST341/sun1.pdf

Under no circumstances should there be so many electrons that an electron shell can form. The incorporation of electrons into an electron shell of the atom only takes place in the deeper layers of the solar atmosphere when the atoms have reached the bottom of the *double layer*. A double layer basically works like a large electrical capacitor, with the sun in the background as an accumulator.

The condensed atoms slowly rain down on the surface, which the astronomer identified as a calcium melt, because the sun, as a star of the spectral class G2, has a high calcium content which corresponds to the Suns supposed mass density of 1.41 g / cm³.

In this solar model there is no reason why only helium should be created first and only when the hydrogen is used up should the helium fuse to form heavier elements. Because of the random distances of the particles, all possible combinations and arrangements of atomic nuclei will occur at the same time. The unstable combinations will transmute until finally the stable combinations remain. Then they meet the electrons that ran ahead and that have collected on the *double layer* between chromosphere and corona. In this way they can fill up their electron shells and the visible signs of this process are the absorption lines of the newly formed atoms.

5.5 Technical Fusion Experiments

The second half of the last century, also known as the Cold War phase, was marked by the use of nuclear energy by splitting uranium atomic nuclei from the deep rock of the earth. In addition to the development of weapons technology, the peaceful use of nuclear energy was promoted and used on an industrial scale. After an initial euphoria, the problem of the final disposal of the spent fuel elements soon arose. These fission products of heavy elements are still radioactive for many centuries, which severely damages organic life. The meltdowns of the reactors in Lucens, Chernobyl and Fukoshima also made the dangers of this technology clear to the world.

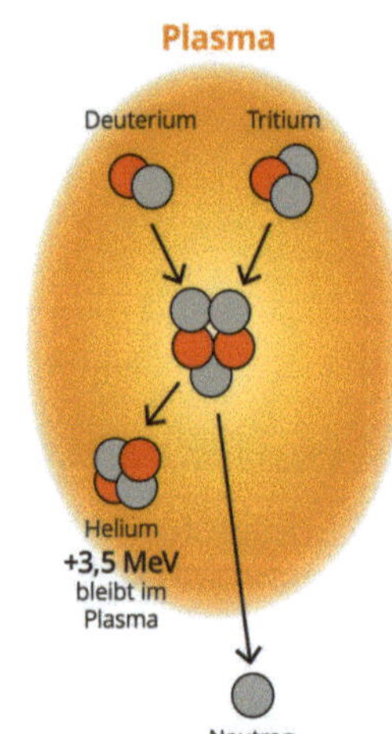

Fig.43: Concept of nuclear fusion as the reverse of nuclear fission

It soon became clear that the sun draws its large amounts of energy from the reverse process, the fusion of atomic nuclei, when

Edward Teller enveloped the atom bomb in a hydrogen coat in 1952 and the world was able to observe the 800-fold explosive force of the Hiroshima bomb.

Nuclear fusion itself does not produce long-lived radioactive waste products like those that occur in nuclear fission, which makes them appear in a more favorable light compared to controlled nuclear fission. Technical experiments for the peaceful use of nuclear fusion are based on experience with hydrogen bombs. While free neutrons are observed during nuclear fission, which release the fission products and keep the fission process going, it was not possible to confirm the role of neutrons in the progress of the fusion. Nevertheless, the belief persists that the neutrons could also produce the fusion effect, as the diagram of the Karlsruher Institute for Technology shows in Fig.43.

Based on the classical notions that the sun is a fiery ball of gas, inside of which hydrogen is compressed so strongly at high pressure and high temperature that it fuses to helium and the heat is transported to the surface by neutrinos, Andrei Sakharov began at the Muscovite Kurchatov Institute with first attempts at technical nuclear fusion. His principle was to enclose a plasma ring in a magnetic field, which should be heated up enough to start the nuclear fusion. The idea was that after ignition, the fusion would continue by itself, as had been observed in nuclear fission through the multiplicative effect of free neutrons. That is why the experiments were designed for pulse operation. This principle was given the name 'tokamak'. It is the abbreviation for "тороидальная камера в магнитных катушках" (toroidalnaja kamera w magnitnych katuschkach) translated toroidal chamber in magnetic coils.

There were identified two problems with this principle. The first was that the Lorentz force swirled the plasma and the chamber walls cooled the plasma. The swirling was finally countered with a complicated magnetic field design that was modeled on a multi-twisted Möbius tape (Fig.44) This created a helical magnetic field. This principle was called the Stellarator, meant a machine which works like a star. However, more than 60 fruitless years had to pass before the new principle was decided.

On December 10, 2015, a short flash of light in the form of a helium plasma was finally generated for the first time on the Tokamak Stellarator Wendelstein 7-X fusion generator in Greifswald, Germany. To carry out this experiment, it took 9 years of construction, in which several hundred tons of material were used. The project devoured over a billion euros.

Fig. 44: Stellarator principle IPP (Max-Planck-Institut für Plasmaphysik) (Neckline)

On June 25, 2018, the Max Planck Institute for Plasma Physics issued a press release:

Stellarator record in fusion product: *„Wendelstein 7-X has now set a record. Because it never reached a previously measured maximum value for the so-called fusion product. This product of ion temperature, plasma density and energy containment time indicates how close one comes to the reactor values for a burning plasma. In the test runs in 2017, the plasma in the reactor was heated to an ion temperature* *of around 40 million degrees and had a density of 0.8 × 10²⁰ particles per cubic meter. "*

Up to 75 megajoules of heating energy were used for the heating and the plasma survived 2 seconds. However, there was still no sign of fusion. If Wendelstein 7-X should ever work, it wouldn't work like a star. Where is the double layer that creates the temperature gradient?

Thermodynamics says that you have to dissipate entropy if you want to create an order. Around every star there is a temperature gradient that dissipates the radiation energy and every star is an anode that first sucks away the surplus electrons. But the stellerator shows neither one nor the other necessary property. The only goal is to heat up the plasma, i.e. to supply entropy and avoid collisions due to the feared cooling. But it is precisely the temperature gradient that is important for the formation of a structure, as can be observed everywhere in nature condensation on cold surfaces.

The working principle of the sun cannot be as complicated as academic nuclear physicists believe. The proponents of the theory of the *Electric Universe* under the direction of Wal Thornhill took the view that sunlight is in principle like the anode light of a gas discharge and asked Montgomery Childs to test this view experimentally. Montgomery Childs[55]) and his small team began in 2013, unnoticed by the public and independently, to recreate Scott's electric solar model in the laboratory by observing the stigmergy of a plasma of hydrogen and nitrogen ions in a vacuum chamber, as occurs interstellar in the spectra of galaxies.

---

55   M. Childs – *The SAFIRE-Project;* https://aureon.ca/history

To generate the plasma, two opposing large plate-shaped electrodes are arranged in the vacuum chamber, between which a small spherical copper anode is situated, which represents the solar model as Fig. 44 shows. The project was named SAFIRE. The name SAFIRE stands for Stellar Atmospheric Function In Regulation Experiment. For this experiment, a lot of parameters, such as gas composition, pressure and temperature, voltage and current have to be regulated. The measurement technology of these parameters must also be closely monitored and controlled. The model has shown that it is possible to keep a stable high-energy plasma.

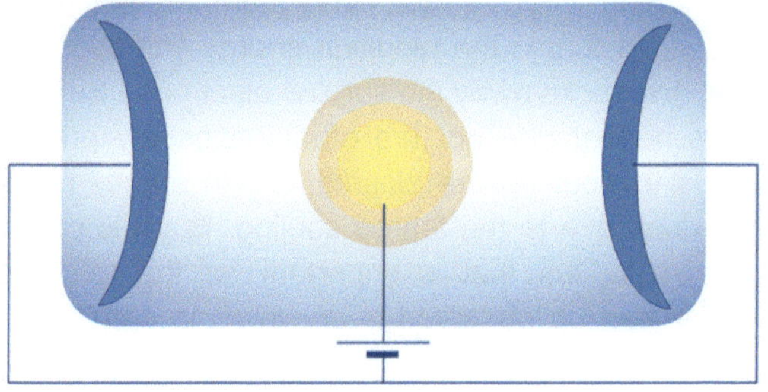

Fig. 45: Scheme of the SAFIRE chamber

Passing through the current-voltage characteristic resulted in several stable levels, in which tufts on the anode first appeared, later double layers and finally a very hot plasma. They vary in intensity and number by a number of interacting factors as well as voltage and current. In this case, operating voltages in a range of 300 to 400 V and a current of 1.5 to 2 amps and a chamber pressure of 20 torr were used.

This experiment had to be carried out very carefully in small steps, as there was always the risk that the entire system could be destroyed as a result of an uncontrolled lightning discharge. With the sun model of the SAFIRE-Project, many of the assump-

tions of Scott's anode model were confirmed. It was also observed that the temperature in the immediate vicinity of the anode first decreased and then rose again sharply with increasing distance.
The stronger the current, the hotter became the plasma and anode. There was a risk that the anode would begin to melt at the predicted power limit. Surprisingly, however, only 7% of the actually expected energy was used and the experiment had to be terminated. Where did all the energy come from?

After the last experiment, the color of the anode surface had changed slightly due to the effects of heat. The anode was then examined for possible deposits under the electron microscope. In fact, tiny droplets of various chemical elements from the plasma condensed on the anode surface, which were actually visible under the electron microscope, chemical elements that were not previously present in the chamber.

Apparently, the structure of the atom begins with the building blocks of deuteron and triteron, which merge from the hydrogen molecule with the release of an electron with one further electron within the plasma to form a deuteron or with an additional proton to form a triteron.

To do this, the molecular hydrogen must be ionized in an electric field and one of its shell electrons must be robbed at the anode. There remain two protons and one shell electron. But this electron will now probably get between the two protons, reduce its vortex, and so the two protons are connected to a magnetic deuteron by releasing radiation energy. If both electrons are lost at the anode, the protons are sent to the cathode or captured by a deuteron. This process corresponds exactly to the first case of Prigogine's theorem about entropy in open systems, where dS_{ext} <0 due to external cooling and dS_{system} <0 due to electron withdrawal. This is exactly what provides the prerequisite for building a higher order from the elementary building blocks. As Childs reported, the necessary electron deficiency is intensified by a catalyst in order to set the fusion in motion. The advantage of this experiment is that the reaction can be stopped immediately by switching off the external direct current. This means that an accident analogous to the meltdown can never occur.

Since the cooling removes more entropy from the system than is generated internally. it is exactly the effect that we want to use technically in nuclear fusion.

Fig.46 shows the periodic table of the elements with the elements found on the anode of SAFIRE marked by yellow fields. All of these elements have also been identified in the Fraunhofer lines of the solar spectrum.

Child's SAFIRE experiment showed that we understand the working principle of the sun's energy production. It is physically simple, but technically demanding in terms of materials. Nevertheless, the path seems technically accessible in order to give hydrogen technology a higher level of efficiency via nuclear fusion in small blocks.

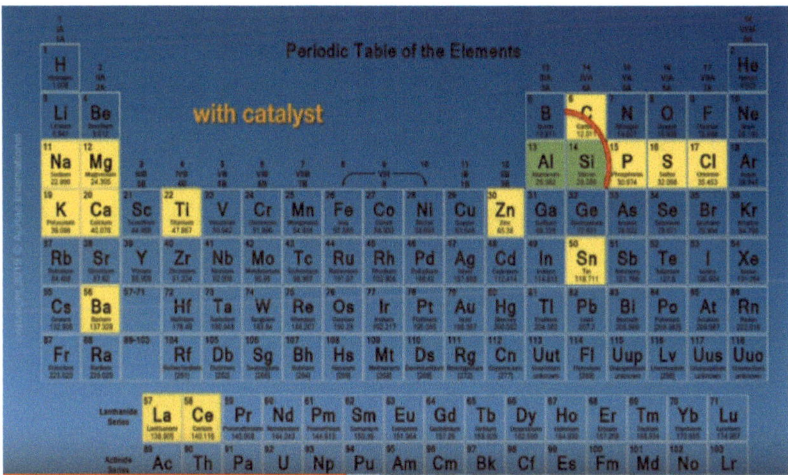

Fig.46: The yellow fields show the founded Elements

6 Questions to which You cannot get Serious Answers

"Science cannot solve the ultimate mystery of nature. And that is because, in the last analysis, we ourselves are part of nature and therefore part of the mystery that we are trying to solve."   -Max Planck.

In 1933 the astrophysicist Edward A. Milne postulated the cosmological principle:

The cosmos is isotropic and homogeneous.

It not only looks the same in all directions, but according to this principle you can assume that cosmological phenomena do not differ from earthly phenomena, which is why they must be based on the same physical pricips on bigger scales. Only then they run much slower then on earth. This opens up the possibility of testing these phenomena in the laboratory. One of the most important physicists who applied this principle was the Swedish Nobel Prize winner Hannes Alfvén.

Nevertheless, dubious businessmen who give themselves a scientific coating sell us sensations from the cosmos, which subsequently lead to the abstruse speculations in public.

I have highlighted the most common of these unanswerable questions here.

6.1 How was the Initial State of the Cosmos?

The Big Bang would be the birth of space and time, claim the followers of this belief. However, space and time are spiritual inventions of man in order to find one's way in the material world. In order to be able to measure a material volume, a relation between volume and imagined space is necessary and in order to be able to measure a movement. An order relation between two movements is necessary, one of which I call time.

The cosmic world order only knows the dynamics of its objects. Humans need relations with which they can compare the dynamics of movement in the physical volume and map them in their consciousness. That is the essence of relativity, the relation between observer and object.

Relativity always only means that the choice of the reference point depends on the observer, but not that the space is curved and also not that the time is independent of the movement, since the speed is defined as the ratio of distance to time interval. Independence of time from the path, on the other hand, would mean that time would be perpendicular to the path. I could move from one place to another without wasting time. However, every analog watch proves that this is not the case, as the arrowhead of the pointer describes a path on the dial that provides information about the past time. So if I want to determine time, I need movement. But what movement has it been since the beginning of the world?

From the viewpoint of thermodynamics you cannot put an explosion at the beginning of all, because an explosion destroys an order, which contradicts the idea of creation. So it is perfectly reasonable to see an act of creation as the beginning. The problem is the creator. Who made the creator? A dispute would immediately break out as to whether it was a self-organization or an external organization.

Both principles interact in open systems. Organizations are created for a purpose and have goals. Nature, however, reveals neither purpose nor goal, which is why the concept of organization is inappropriate in connection with physics. I have therefore chosen the term stigmergy.

On the other hand, there must be a connection, a stigmergy, between objects of the same type, which must stand out against a background if one wants to recognize an ordered structure. Consequently, no structure can emerge from nothing, as the representatives of the Big Bang would have us believe.

The idea of running time backwards to get to a beginning is useful for a sequence of images. Images are independent of time and do not interact with one another. However, a development process is not a sequence of images. Each new development step is dependent on a previous development step in accordance with the interaction between the individual and the environment, and there are different options for decision-making at each stage of development. So the reconstructing viewer is constantly faced with the question: Where does this development come from? Such a reconstruction may still be possible if there are still traces of developmental steps as in the case of stars and galaxies.

But what if such traces are unknown or completely missing, as is the case with larger structures? There will be then no scientific answer to this.

6.2 Is the Space expanding?

If the concept of space serves as a mathematical order relation, it is not helpful to assume that it is continuously changing. My yardstick should be constant. If I increase my scale, it means scaling in the sense of zoom out. But that has nothing to do with physics.

But this is about another phenomenon that is being misinterpreted. Due to the spectral redshift, which Lemaître and the representatives of the Big Bang interpret as the Doppler effect, it gives the impression that all radiation sources are moving away from each other, with very few exceptions. As we have already established, the spectral redshift is based only to a small extent on the Doppler effect. The greater part is due to the reduction in light energy through its absorption in the cosmic medium in the form of dust between the light source and the observer. This makes the spectral redshift unsuitable for the reliable determination of the distances in the cosmos, as Halton Arp has already established.

We must then ask whether the closed volume is expanding. However, the volume is related to the mass. We would have to look for a source of mass in the cosmos, and since the effect of expansion appears to come from the earth, the earth itself would be the source of the mass. But Hermann von Helmholtz discovered a long time ago that mass and energy were preserved. In 2011 the Nobel Committee disregarded all the fundamentals of physics and awarded a prize for the accelerated expansion of a closed cosmos. But the real cosmos is open one. Try to blow up a balloon with an input hole and an output hole.

The answer to the next two questions is also related to this misinterpretation.

6.3 How big is the Cosmos and what is behind?

The question assumes that the cosmos is closed. In this point I share Pope Francis' view that, contrary to earlier beliefs, the cosmos is open and that limits are only set by our observation horizon. What lies beyond those boundaries is a matter of belief. After the redshift of the spectra has proven to be an unusable method of determining distance, we only have the chance to observe the starry sky with space probes at a great distance from the sun, in order to obtain stereometric recordings that we can evaluate for determining the distance. To determine the distance of an object, we always need two images from different viewpoints. We know the length of the baseline that is bounded by our observation points and we can determine the angle between the other observation point and the observed object. Then you can calculate the distance of the object using trigonometry. So far, the largest possible baseline has been the earth orbit diameter. From this we derived the parsec as a measure, which corresponds to about 3 light years. Astronomers believe that they have found the limit of our present-day observability by means of redshift at a distance of about 15 billion parsecs. However, the information does not contain any measurement errors and is therefore speculative and perhaps far too optimistic.

 The observability of the cosmos will always remain finite, which separates it from the infinite universe. What lies beyond the observation horizon is in any case unfathomable and the subject of unscientific speculations.

Whether there are parallel worlds that can be reached via wormholes is pure fantasy and lacks any physical basis. The fact that well-known scientists are dealing with it testifies to the decline of astrophysics as a serious science.

6.4 How old is the Cosmos?

The age of the cosmos is given in the Big Bang model as 13.8 billion years. It is written by Wikipedia:

> »After an initial accelerated expansion called the inflationary epoch at around 10^{-32} seconds, and the separation of the four known fundamental forces, the universe gradually cooled and continued to expand, allowing the first subatomic particles and simple atoms to form«

Expansion from nothing and why force separation? If there is heat, there must be charged matter in movement and what force slows down the expansion if there is no mass at all? Have we not learned that expansion has to do with warming the mass?

The creationists put the age of the cosmos at about 6 thousand years according to the biblical family tree from Adam on. Both models build on the power of God's words "Fiat lux" (et facta est lux). Both models ignore the fact that light is the result of an electrical discharge in matter only the former sounds more complicade.

Is the former statement more scientific than the latter? Both statements are based on speculation without physical base, although we are inclined to have more confidence in the first statement because we have already collected more evidence of the history of the earth than of the history of the cosmos in our museums and we consider the earth with the sun to be younger than the cosmos, since sun and earth move in it.

We have made the experience of the basic building blocks of matter, the electrons and protons, that they are stable over unimaginable periods of time. This means that these building blocks exist outside of any time. There are also stable elements made from these two building blocks. There is no clock with which one could ever determine the age of the cosmos. It is different with galaxies. There you can find at least a measure of time. If you can determine the concentration of hydrogen in the galaxies, it turns out that spiral galaxies must be younger than structure-less galaxy clusters. Spiral galaxies have strongly developed hydrogen lines that decrease with age and with increasing core size. So at least we can create an age-related order among the galaxies, but without being able to establish a reference to other cosmic events, because we do not know when the light that we are observing today went on its journey through the cosmos. .

6.5 Can we leave our Planet Earth?

The successful Apollo 11 mission to the moon nourishes the desire to visit other planets as well. Even the permanent settlement of these planets is being discussed.

It reminds me of my first experience with a small aquarium. In winter my aquarium had to move to a darker place by the stove. That didn't benefit the planting. They died and not much later the fish. The closed system overturned.

The establishment of a biosphere is subject to the law of thermodynamics, as discussed in Section 1.2. Even if we succeed in maintaining a closed biosphere on earth for some time, it is doubtful whether it can function under space conditions, since there is either less energy available there or too much energy that cannot be dissipated.

We are just realizing that at the beginning of the 21st century we are about to destroy our huge planet because we are demanding more resources from it than it can permanently provide us
or to put it thermodynamically, because we are generating more entropy than the earth can dissipate.

How long should be able to survive a small biosphere populated with humans in the form of a spaceship, where with the speed of the solar wind we have to calculate travel times of hundreds of generations before we reach a neighboring planetary system with an orbit that is worth living in for humans?

What should the well-dosed energy supply on this long space journey if it is not an artificial sun that would have to be built into the spaceship?

So space tourism is a highly questionable and resource-consuming matter which will demand more sacrifices from humanity than it will bring to its benefit.

**Our spaceship is the planet Earth.
We cannot get a more comfortable spaceship.**

Our earthly society is also subject to cosmic dynamics. The metabolism to preserve them takes place with raw materials and private capital. This creates added value in capital on the one hand and garbage on the other. Nature has to turn society's metabolic products back into usable raw materials. Man-made global warming is an expression of the fact that nature can no longer cope with this task. The increase in CO_2 is only one measurable indicator. That does not seem to be general knowledge of society yet. I cannot yet see any global action plans.

Epilogue

The ideas and theses developed in this book differ considerably from the current academic doctrines of modern physics. Yes, they even question the math-based textbooks to a large extent, which is why I have to say something about the background to their creation.

The mathematician David Hilbert should once made the arrogant statement in 1912:

Physics is much too hard for physicists [56])

When I complained during my studies that I couldn't develop an image of things and processes, especially in so-called modern physics, the answer I usually received, was that it wasn't necessary. I just have to calculate. That eventually stopped me from considering theoretical physics as my goal in life. My interest then turned more to the analytical measurement techniques of atoms and molecules and the evaluations of the measurement protocols led me to computer science and then increasingly to the study of development of mathematics. It wasn't until my working life was over that I found the time and muse to reflect on the riddles of my student days.

---

56   C. Reid – *Hilbert;* Springer-Verlag, Berlin/Heidelberg/New York 1970, S. 127, books.google.de https://books.google.de/books?id=RtnvBgAAQBAJ&pg=PA127&dq=%22is+much+too+hard+for+physicists.%22

So if the knowledge of nature is too difficult for a natural scientist and he has to leave it to the humanities scholars in order to axiomatize it, then something has gone completely wrong.

Since my professional retirement in 2005, I had set myself the goal of really wanting to understand relativity and quantum mechanics. But I don't gain understanding by simply following the thought processes of a thought leader. So I am immediately contained within the limits of the thought leader and that contradicts the principle of this book. I have to cross the borders without the risk of speculation. This is the only way I can escape the paradoxes the theorists have amassed.

Since my professional development has led me from physics to computer science via the study of non-numeric mathematics, I got a much deeper understanding of mathematics than physicists ever got in their training. The practical application of this knowledge in engineering has always been in the foreground. I have only recognized what is real knowledge when it comes to what has proven to be practical. Of course, the human mind can play excellently with mathematics, but nature does not follow generalizations, not transformations and whatever else a lively mind can come up with. Above all, this spirit thrives on idealizations that fall on the engineer's feet if he unconditionally follows the promises of theories when they are put into practice.

So these are the conditions under which a theory emerged that needs to be tested. With mathematics you can prove mathematical calculations, but not scientific relationships. As a student, I could not have this critical look at the mathematical methods and their requirements that physics applied, because there was simply no practical life experience. And that is also the problem of the theorists in the educational establishments.

In German there is the beautiful word *"begreifen"* for understand. It means grasping with your hands. Mathematical formulas are of little help and exceeding the limits of our sensual abilities inevitably leads astray. Immanuel Kant already communicated this to us with his Critique of Pure Reason, but it was not understood if read at all.

The fear of keeping academic teaching clean and the lack of basic philosophical and physical knowledge have given us the many paradoxes of physics over the past hundred years. Whether the paradoxes and the climate crisis will be overcome in the next generation is a question I don't dare to answer. Hope and doubt are balanced.

Acknowledgment

After my first book was published, I had a lively exchange of ideas with a number of scientists. These thoughts stimulated me differently. At this point I would like to mention the names to whom I am grateful for their suggestions, even if we sometimes had diametrically different opinions. Dissent is the strongest motive for formulating your own ideas.

I would like to thank for the fruitful cooperation: Hannes Täger, Klaus Gebler, Jean de Climont, Andre Koch Torres Assis, Vladimir Netchitailo, Christian Joos and, last but not least, the Thunderbolt community under the direction of Wal Thornhill with the contact via Susan Schirott

Of course, I also thank my wife Marlies Hüfner for her patience with me when I was mentally absent despite my physical presence.

Term list

About the author

The author studied physics in Leipzig from 1964 to 1970. He graduated from the Institute for Radioactive Isotopes. He then worked until 1978 at Carl Zeiss Jena in the Analytical measurement technology department on the development of optical measuring devices and software for the analysis of spectral data.

He later took up a position as an assistant at the former technology section of the Friedrich Schiller University in Jena in the field of cybernetics and experimental planning, studied non-numerical mathematics, computer science and other engineering sciences and received his doctorate in this area in 1983.

After the social change in what was then the GDR, the author started a freelance teaching activity in the field of computer science after several Additional qualifications, which he carried out until he reached retirement age in 2008.

Only then did he begin to work again as an independent scientist in physics and establish contacts around the world. Halton Arp and Paul Marmet, Benoit Mandelbrot and Gerald Pollack in particular had a lasting influence on the author's thinking. So he as a volunteer finally found his way to the Thunderbolts, a small avant-garde community of scientists and engineers who are working on a new understanding of the cosmos. Even if not all the ideas represented there, some of which have a mythological background, are fruitful, the author, with his basic understanding of physics acquired in Leipzig, was able to select the fruitful ideas from the sterile ones.

In 2019 and 2020 he published his first book in English and German with the title *Modern Astrophysics Meets Engineering*[57]). He also maintains a website http://mugglebibliothek with blogs and YouTube movies on the nonacademic paradigm of the Electric Universe.

---

57 https://www.bod.de/buchshop/modern-astrophysics-meets-engineering-mathias-huefner-9783751920186